International Relations in
Uncommon Places

International Relations in Uncommon Places

Indigeneity, Cosmology, and the Limits of International Theory

J. Marshall Beier

First published in 2005 by
PALGRAVE MACMILLAN™
175 Fifth Avenue, New York, N.Y. 10010 and
Houndmills, Basingstoke, Hampshire, England RG21 6XS
Companies and representatives throughout the world.

PALGRAVE MACMILLAN is the global academic imprint of the Palgrave Macmillan division of St. Martin's Press, LLC and of Palgrave Macmillan Ltd. Macmillan® is a registered trademark in the United States, United Kingdom and other countries. Palgrave is a registered trademark in the European Union and other countries.

ISBN 1–4039–6902–7

Library of Congress Cataloging-in-Publication Data

Beier, J. Marshall.
    International relations in uncommon places : indigeneity, cosmology, and the limits of international theory / J. Marshall Beier.
        p. cm.
    Includes bibliographical references and index.
    ISBN 1–4039–6902–7
        1. Indians of North America—Government relations. 2. Indian cosmology—North America. 3. Indians of North America x Politics and government. 4. Hegemony—North America. 5. Cultural relations. 6. International relations. 7. North America—Race relations. 8. North America—Relations. 9. North America—Politics and government. I. Title.

E91.B45 2005
323.1197—dc22   2004061825

A catalogue record for this book is available from the British Library.

Design by Newgen Imaging Systems (P) Ltd., Chennai, India.

First edition: June 2005
10 9 8 7 6 5 4 3 2 1
Printed in the United States of America.

For April

Contents

Acknowledgments

A considerable debt of gratitude is owed to the many people who have given me their support and encouragement through the course of researching and writing this book. For their generosity with their time, their patience with my questions, and, above all, for extending their trust, I am deeply indebted to all those from South Dakota without whose input the book would not have been possible. Under very trying circumstances in 2000, members of the Grass Roots *Oyate* took time from more pressing matters to explain not only their situation but its broader context as well. Others have been equally supportive, sharing their thoughts and perspectives on a range of key issues and ideas. I am especially grateful to Charmaine White Face for her invaluable guidance, supportive feedback, and for her willingness to have her personal experiences inform this project. At York University (where I first asked why Indigenous peoples had been overlooked by International Relations), Robert Albritton, David Mutimer, and Sandra Whitworth contributed immeasurably both through their readiness to entertain my ideas and with the important suggestions they made from the earliest stages of my research. Mario Blaser, David Campbell, Peter Penz, and Simon Philpott each provided engaging feedback on the whole of the manuscript. A number of people read and gave thoughtful comments on various chapters: Anna Agathangelou, Samantha Arnold, Shampa Biswas, Geeta Chowdhry, Roxanne Dunbar Ortiz, Hugh Gusterson, Chris Hendershot, Tami Amanda Jacoby, Lily Ling, Cristina Masters, Heather McKeen-Edwards, Rabea Murtaza, Sheila Nair, Peter Nyers, Randolph Persaud, Alina Sajed, and Virginia Tilly; their comments and those of two anonymous reviewers have made this a better book.

The Centre for International and Security Studies at York provided generous support of the initial stages of my research and I benefited greatly from the collegial environment fostered there, in particular by Joan Broussard, Heather Chestnutt, Ann Denholm Crosby, David Dewitt, Steven Mataija, and others too numerous to list here. Many engaging conversations with colleagues in the Political Science Department at McMaster University helped me to think through key issues and ideas. Will Coleman, Robert O'Brien, Tony Porter, and Richard Stubbs all shared invaluable advice about proposing and preparing the manuscript. At Palgrave, Steven Kennedy saw to it that my proposal found its way to the appropriate desk and the consummate professionalism of Toby Wahl and Heather Van Dusen took all of the anxiety out of seeing the manuscript through to publication.

In writing this book I have drawn substantially on two earlier published works. I am grateful to Routledge for permission to include a revised version of my "Beyond Hegemonic State(ment)s of Nature: Indigenous Knowledge and Non-State Possibilities in International Relations," published in 2002 in Geeta Chowdhry and Sheila Nair, eds., *Power, Postcolonialism and International Relations: Reading Race, Gender and Class*. I am equally grateful to Nova Science Publishers for their permission to use a revised version of my "Of Cupboards and Shelves: Imperialism, Objectification and the Fixing of Parameters on Native North Americans in Popular Culture," published in 1999 in James N. Brown and Patricia M. Sant, eds., *Indigeneity: Construction and Re/Presentation*.

My family has been interested and supportive throughout and, I know, will be no less pleased than I to see the volume come to fruition. Finally, I am eternally grateful to April Spencer, who made many personal sacrifices because she also believed in the importance of this project, and to Kaelyn Beier in whom we both believe.

J. Marshall Beier
Hamilton, Ontario
Canada
September 2004

Introduction

I n January 1997 I attended a conference at McGill University convened to consider the long-awaited report of Canada's Royal Commission on Aboriginal Peoples. Released two months earlier, the five volume, 3,200 page report marked the culmination of a more than five year intensive inquiry into the relationship between Canada and its First Nations—a process begun in the aftermath of the so-called Indian Summer of 1990 when a violent confrontation between police and Mohawk protestors at Oka, Quebec precipitated a 78-day crisis that escalated to the deployment of military units.[1] The conference was well attended by academics (mainly specialists in Canadian politics and First Nations' issues from a range of disciplines), prominent members of the Canadian establishment (a former prime minister among them), and, of course, Aboriginal people (including elders, activists, and other community leaders). Having made my disciplinary "home" in International Relations,[2] I might have seemed a little out of place as, to the best of my knowledge, I was the only one from my field in attendance. Indeed, a few of the academics I met did betray some surprise upon learning of my disciplinary affiliation. But I also met Anishnabek, Cree, and Mohawk people, none of whom seemed to give it a second thought.

What had brought me to the conference was a sense that what took place at Oka some six-and-a-half years earlier counted as international relations even if it did not seem to count to International Relations. Though I arrived with a fairly strong sense of the difficulty of trying to talk about Indigenous[3] peoples in International Relations, I still had only the vaguest inkling of the significance of conference participants' differing reactions to my disciplinary affiliation—in the end, the most instructive aspect of the conference in light of the project that was to evolve. Indeed, the disjuncture I noted then is revealing of the starting problematic of this volume: specifically, the near complete neglect of Indigenous peoples by International Relations scholars.

The exceptions here are noteworthy,[4] but will not be central to my focus in what follows for the simple reason that they are just that: exceptions. Notwithstanding the important insights to be had from them, all remain quite decidedly relegated to the margins of International Relations and have made precious few inroads into the research programs and curricula of the discipline.[5] Heeding Steve Smith's perceptive advice that silences are a discipline's loudest voices (Smith 1995: 2), it is this circumstance that I take as my point of departure.

How do we account for this striking omission by disciplinary International Relations even as we reach the end of the United Nations' International Decade of the World's Indigenous Peoples? Certainly, nothing on the agenda at the McGill conference could account for it. Much to the contrary, the presentations and discussions suggested almost no end of senses in which the issues under consideration were issues of international relations: the legacies of conquest and enduring colonialism in the Americas were considered; different conceptions and competing claims of sovereignty were weighed; some participants reflected upon the history of Indigenous peoples' involvement at the United Nations and in other international fora; and, Kahnawake newspaper editor Kenneth Deer recounted the details of the Two Row Wampum as a lasting record of an early-seventeenth-century treaty between the Haudenosaunee Confederacy and the first Europeans to arrive in their lands. The Two Row Wampum, so named for its two rows of purple beads against a white background, symbolized the idea that the Haudenosaunee and the European newcomers would coexist in friendship, but with equality and autonomy of beliefs and lifeways. And this suggests that a starting point in addressing International Relations' inattention to Indigenous peoples might be to ask how it is that this agreement failed. A first, tentative answer, of course, is that the European view of appropriate international practices in the Americas ultimately turned out not to be coexistence or even diplomacy (except, from time to time, as an expedient). Europeans opted instead for conquest.

My broad purpose in this book is to explore the ways in which International Relations has internalized many of the enabling narratives of colonialism in the Americas, evinced most tellingly in its failure to take notice of Indigenous peoples. More particularly, International Relations is read as a conduit for what I am calling the "hegemonologue" of the dominating society:[6] a knowing hegemonic Western voice that, owing to its universalist pretensions, speaks its knowledges to the exclusion of all others. That the hegemonologue so often goes as if unheard underscores that it bears our "common senses," those things we know so well about the world that it would be unthinkable to subject them to critical scrutiny; that we might do so seems not even to occur to us. What this signals is that our existing critiques of various onto-logical and epistemological commitments and positions do not reach all of our knowledges because they do not extend to an interrogation of cosmologies whence the foundations of these knowledges can be traced. The cosmological commitments of the dominating society are not generalizable to the whole of humanity—indeed, they sometimes stand in rather stark contrast with those of Indigenous derivation. But because they are spoken over and against all others through the impositions of the hegemonologue, they also effect violences of erasure. Disenabling the values and commitments upon which Indigenous peoples' self-knowledges—and, therefore, resistances—might be predicated, they ideationally undergird the contemporary European settler states of the Americas and elsewhere. For this they are inseparable from the advanced form of colonialism that is politico-normative heir to the original project of European colonial conquest and domination. And for speaking the hegemonologue and participating in the reproduction of its attendant knowl-edges, International Relations is likewise identifiable as advanced colonial practice.

Though I inquire briefly into empirical contexts involving several different Indigenous peoples of the Americas, my principal focus is the Lakota people of the Northern Great Plains of North America.[7] An examination of tradi-tional Lakota cosmological commitments and the distinctive conceptions of things like security and the good life that derive from them is instructive, not only because they have been and continue to be particularly silenced, but also because the people to which they belong have been so marginalized. The terms and details of this peripheralization ought to be of special interest to students of International Relations to the extent that they involve the "unmaking" of a once recognized "sovereign" unit in the ostensibly "inter-national" system. As part of a stateless Indigenous people, contemporary Lakota traditionalists fall quite decidedly beyond the range of the conceptual lenses of the theoretical mainstream of International Relations. But, as we shall see, the full range of contending approaches to theorizing the interna-tional is inextricably bound up in the reproduction and reinsinuation of hegemonic cosmological commitments and, by extension, is complicit in the denial of Lakota cosmology as well. We are thus confronted with a very clear perspective on the culpability of International Relations in the ongoing project of advanced colonialism.

Interestingly, a traditional Lakota worldview leaves open the possibility of cosmological diversity without giving rise to troubling uncertainties about what can be "known." Though defining its particulars exceeds the bounds of the present project, we may nevertheless infer from this an admonition to engage Lakota knowledges and ways of knowing as constitutive of new (to International Relations) and potentially counter-hegemonic bases for theory. The challenge, then, will be to *listen* to Lakota voices, not simply to appropriate them. And this will require considerable reflection upon unenunciated and seldom acknowledged fundamental assumptions and understandings of the universe—in short, a critical examination of cosmological commitments.

To the extent, however, that I argue that we can never entirely step outside of who and where we are, it is necessary that the reader know something about me and my motives in writing this book. My first aim is to contribute to resistance against the violences of advanced colonialism. But this calls for an important caveat: as a non-Indigenous person, my intention here is not to raise the "authoritative" voice of the academic as if to validate the myriad Indigenous voices that already have been brought to bear on many of the issues I consider herein. The latter need no such endorsement and, were this not the case, may as readily find it in the works of a good many noted Indigenous scholars. As yet, however, these voices have scarcely been heard in International Relations, and so it is to those with whom I share disciplinary affiliation that I most purposively direct what follows. Striving to remain conscious of the position of material privilege afforded by my position in the academy, I am also keenly aware of the less often acknowledged discursive privilege that accompanies teaching, writing, and publishing. It seems fit that this particular privilege should also be exercised introspectively in calling on others working in International Relations to consider the part that we have all played in sustaining the violences of advanced colonialism. Though I do not presume to speak on behalf or in place of Indigenous voices—indeed, this would violate a central commitment of the project—I do hope to ally with them in emancipatory undertakings that, for reasons that will become clear, are not for me to specify.

I venture rather more confidently to articulate the terms of emancipation in respect of another referent: International Relations which, as we shall see, has paid quite a considerable price for its implication in the ideational dimension of colonialism/advanced colonialism. For the same reasons that international theory manifests as an advanced colonial practice, I argue, it has limited our imaginings of the international and subverted the development of a broadly emancipatory project. And while I generally find it most fruitful to discuss International Relations more broadly, I am also interested in the implications of all of this for our ability to effectively engage in the more focused study of security. Here too, the violences of advanced colonialism are readily found at work, denying the legitimacy—even the plausibility—of the self-defined security concerns of Lakota traditionalists. Indeed, the varied ways in which we are used to thinking about making ourselves "secure" are, in some particulars, identifiable as sources of insecurity for Indigenous people(s). Our disciplinary securing of the boundaries of "security," pronouncing upon its content and referents, involves erasures that work to sustain advanced colonialism. At the same time, delimiting this and other concepts causes us to miss seeing that there may be a much richer field of possibilities to inform our theorizing than those we have imagined. In seeking to uncover the ways in which International Relations is implicated in advanced colonialism, then, I hope to make a contribution toward freeing it from the limiting effects of these complicities.

Finally, I am motivated, at least in part, by my discovery of myself as an "nth" generation colonizer. I grew up a Canadian of European descent, privileged socially and economically. I know that my ancestry includes forebears on the leading edge of the colonial encounter in what is now the northeastern United States. And centuries later I make my home on land made available for my habitation by a full range of the violences set forth from the arrival of Europeans in the Americas: like the neighboring Huron, the original Attiwandaron inhabitants were destroyed as a people in a war with the Haudenosaunee Confederacy that owed much to the disruptive influences of the arrival of Europeans, and the land subsequently changed hands several times, passing to the Mississaugas, then to peoples of the by-then broken Haudenosaunee Confederacy, and eventually to European settlement. Even knowing this, I cannot shed the colonial inscriptions I bear by simply relocating; spatial adjustments are no solution because I was born here and also because participation in colonialism can be just as acute from afar.[8] In any event, my (post)colonial identity owes as well to the privileged social site I occupy. What I can do, however, is to answer my participation and complicity in enduring structures of colonial domination with a self-conscious resistance that seeks to lay bare the hidden workings of colonialism in my scholarship and, no less, my everyday life. Still, there is a degree to which I will always be implicated—I thus find that I am myself a postcolonial subject, to the extent that I am indelibly inscribed by colonialism/advanced colonialism.

None of this is proffered gratuitously. Indeed, I might quite readily agree with any suggestion that I have not provided nearly enough detail about myself here. Even so, what I have offered might offend the sensibilities of those who would have us remove ourselves from our work, leaving "truth" or "knowledge" to stand apart from our comparatively unseemly subjectivities. For my own part, however, I give no deference to the positivist-inspired epistemological commitments that would insist upon the radical separation of a containable subject and a contained object. Rather, I am inclined toward the view that, inasmuch as it necessarily affects what I am apt to see and how I might begin to understand it, who/what I am is every bit as relevant as who/what I research and write about. To pretend otherwise would be to disavow my disavowal of positivist pretensions to having effected an unproblematic separation of subject and object. Moreover, it would be to proceed as though my privileged position(s) in sociopolitical time and space do not bear on my work. If I am necessarily a (post)colonial subject, then all of what follows is quite properly read as a (post)colonial text, equally inseparable from the workings of advanced colonialism. Knowing something about the context(s) of its author, then, will aid the reader in applying the methodological device described by Edward Said as "strategic location"—that is, locating the author in the text and in relation to its subject matters (Said 1979: 20). But there is, perhaps, an even more compelling reason to self-identify: as Katherine Shanley points out, the problem of appropriation of Indigenous voices means that in order to establish their authority to speak, "Indian writers are forced to reveal details of their personal lives and histories in ways few other writers are expected to do" (Shanley 1997: 692). The authority of the voice of the academic is, in contrast, conferred a priori by credentials and by a place at the lectern. Here self-identification becomes important as a means by which to destabilize that authority and situate it socially so that its pretensions do not exceed its competencies. And so I identify myself as many things, among them an "nth" generation colonist. What follows is, therefore, necessarily much less an exposition of what I have *learned* than how I have *understood* it.

The reader is also cautioned that the hegemonologue can be found working through all of what follows in ways that might not always be immediately apparent. A characteristic feature of Western cosmology is constituted by its linear expressions of being. This extends also to ways of knowing as, for example, in processes of inductive or deductive reasoning. This linearity contradicts the circular expressions of being and knowing borne in traditional Lakota cosmology and is therefore something that I work to unsettle in a number of instances. Ironically, however, I have set about organizing this book in the decidedly Western manner of constructing a linear argument, with each chapter set in planned sequence in a progression toward a few points that I would have them support. Breaking with this, I have found, is not easily done if I hope for my argument and the insights that flow from it to be audible to Western disciplinary ears, particularly in disciplinary International Relations. This is, after all, our accustomed way of coming to knowledge.[9] With this in mind, I turn now to briefly trace the course I take in this strategic linearity.

Chapter 1 serves a dual purpose, elaborating the workings of the hegemonologue and laying out my own theoretical commitments. As to the first, I prepare the ground for more focused discussions of the myriad ways in which the hegemonologue works through us and, more particularly, through international theory, disciplinarity, and even the hierarchical-structural arrangements of the academy itself. It is not International Relations alone that is touched by this critique, and so I range across several disciplines in the chapters that follow. My theoretical commitments form the basis for how I propose to "answer" the hegemonologue, but the special challenges of talking about Indigenous peoples in the context of a discipline that remains insufficiently attentive to its implication in advanced colonialism necessitates a more careful elaboration of these commitments than might otherwise be called for. Relying heavily on Derridean deconstructionist strategies of critique, I review some of the objections raised against poststructuralism with the aim of separating important caveats from "red herrings." Still, left with certain reservations of my own, I argue that postcolonial theory, infused through an ethics of responsibility, furnishes the needed corrective.

Chapter 2 introduces important methodological questions as well as some of the barriers to our identifying and engaging them. Discussion here turns on the related problems of disciplinarity and the (re)production of authoritative voice that, together, have conspired to leave students and scholars of International Relations decidedly ill-prepared for the sort of ethnographic research and writing they are increasingly undertaking. Moreover, the disciplinary parceling off of ostensibly discrete knowledge realms has had the effect of disembodying such cross-disciplinary appropriations as have, from time to time, been made—a circumstance that, it is argued, has had deleterious consequences of its own and is analogous to cross-cultural appropriations of knowledge.

Proceeding from insights revealed in chapter 2 and informed by the theoretical commitments outlined in chapter 1, chapter 3 cautions those working in International Relations against presuming that ethnographic research strategies developed in Anthropology and elsewhere are a sufficient basis for the competencies requisite to speak on behalf of their subjects—a point reinforced in an extended caveat on customary guidelines for "ethical" research practices involving human participants. These various threads come together in the call for a methodological commitment to conversation: that is, a strategy that refuses the pretension to appropriate the voice of the Other, working instead toward its audibility alongside our own.

This is a commitment that I have tried to abide by in chapter 4. In outlining the rudiments of a worldview and lifeways (historical and contemporary) fashioned in deference to traditional Lakota cosmological commitments I have worked to make central the voices of Lakota traditionalists themselves. And though I have consulted directly with Lakota people, I specifically disavow any inclination to speak from a knowing subject position of my own. I have, in other words, explicitly rejected participant observation through fieldwork as a means by which to imbue my own voice with authority—the only authority here resides in those voices that I have endeavored to make audible. Also emphasized in this chapter are the continuities of the colonial encounter, from the wars of the nineteenth century, through the early reservation period, and into our present era of advanced colonialism. A persistent theme involves the violences visited not only upon Lakota bodies (genocide) but upon lifeways, culture, and cosmology as well (ethnocide). Against this backdrop, some pressing Lakota security concerns are briefly considered.

The focus of chapter 5 is on violent knowledges—in particular, those articulated through (re)presentations of Indigenous people(s) in the dominating society. Accordingly, I examine the ways in which popular culture (re)presentations of Indigenous North Americans impart knowledges that, rendering them as spectacle, contribute to their consummate objectification. This, in turn, promotes the confinement of Indigenous people(s) to a limited range of temporal and spatial contexts in the popular imaginary. Comparing nineteenth-century forms (and their enablement of the colonial conquest) with examples from contemporary pop-culture, I argue that these (re)presentations fulfill a vital (if not necessarily instrumental) role in the maintenance of advanced colonialism and of the North American settler states through which it is most conspicuously expressed. Having thus highlighted the link between particular knowledges and advanced colonialism, the ground is prepared for chapter 6 wherein I find many of the same knowledges (and their consequences) both underwriting and (re)produced by orthodox social theory. Ultimately, the voice of the hegemonologue is found to be audible through popular culture and orthodox social theory alike in ways that render traditional Lakota knowledges and lifeways implausible and suggest how it is that international theory might be identifiable as an advanced colonial practice.

In chapter 7 I take up the problem of emancipatory violences: the unintended complicities of *critical* International Relations theory with ongoing processes of advanced colonialism in the Americas. In particular, the universalizing tendencies of Western cosmology are shown to be at work in aspects of emancipatory social theory. The result is that the terms of emancipation tend to be dictated from a position of relative privilege and, to the extent that they conflict with Indigenous people(s)' notions of the good life and how best to attain it, this gives rise to violent erasures of its own. As with the orthodoxy, then, I argue that here too the predominance and persistence of Western cosmological commitments serve to invalidate Lakota knowledges and ways of knowing upon which resistances are predicated. This, of course, speaks back to my earlier call for a conversational strategy of engagement with our Others. Though we do not easily escape the hegemonologue, conversation promises at least the hope of unsettling it. And in discovering ourselves (to say nothing of our discipline) as postcolonial subjects in our own right, we may also find that we have more at stake in this than we might have imagined.

Notes

1. On March 11, 1990, following a ruling authorizing a golf course expansion that threatened a Mohawk burial site, Kanesatake Mohawk protesters erected barricades to block construction. Two months later, at the request of the mayor of the town of Oka, Quebec provincial police raided the barricades, precipitating an exchange of gunfire that left one police officer dead. A second line of confrontation opened when Kahnawake Mohawks, reacting to the police assault at Oka, prevented traffic from crossing the Mercier Bridge near the Montreal suburb of Chateauguay. After violent clashes between police and angry crowds trying to force the reopening of the bridge, more than 2,500 soldiers were deployed to Chateauguay and Oka. Shortly thereafter, an agreement was reached to end the standoff at the Mercier Bridge; army advances overtook the barricades at Oka on September 1 and forced the surrender of the last of the protesters nearly four weeks later. Thirty-four protesters who faced various criminal charges stemming from their roles in the standoff were later acquitted and a Quebec coroner's report was highly critical of the initial police assault. The disputed land was purchased by the federal government in 1997 and put under Mohawk control, though not in deed.
2. In order to distinguish between them, I variously use "International Relations" and "international relations" to denote the discipline and its subject matter respectively.
3. I use the terms "Indigenous" and "aboriginal" in different and very specific ways. The former, I use in the manner of a proper noun in reference to the original human inhabitants of the Americas. This treatment is in contradistinction to the more generalized "indigenous" that is more susceptible of appropriation by American-born persons of Euro-American descent seeking to undermine claims to sovereignty by Indigenous people(s). In contrast, I am using "aboriginal" as an adjective because, being more explicitly connected to the pre-Columbian past, it is less ambiguous in reference to "aboriginal warfare" or the "aboriginal condition of Indigenous peoples." "Native" is synonymous with "Indigenous," though I have tried to avoid using it for the sometimes pejorative connotations it has in colonial discourse.
4. These include books by Franke Wilmer (1993) and Paul Keal (2003). Wilmer is also the author of "Indigenous Peoples, Marginal Sites, and the Changing Context of World Politics" (Wilmer 1996). Roger Epp (2000) and Karena Shaw (2002)

identify some of the problems, promises, and prospects for exploring intersections between indigeneity and IR. For a reading of the Great Law of Peace of the Haudenosaunee Confederacy as a security regime, see Crawford (1994). A response to Crawford that disputes this reading is Bedford and Workman (1997).

5. See Wilmer et al. (1994).

6. I prefer this term to the more conventional "dominant society" that, it seems to me, is inappropriately suggestive of an objective and static relationship. The term "dominating society" is thus intended to reflect the ongoing processes of reproduction of the particular knowledges and material conditions requisite to sustained advanced colonial domination.

7. The Lakota, the Teton division of the Dakota people, may be better known by the name "Sioux," usually understood to comprise the aggregate of all Dakota peoples.

8. See, e.g., the discussion of German "hobbyism" in chapter 7.

9. For an example of an argument constructed in a way that begins to break out of linear ways of coming to knowledge, see Der Derian (1997). As effective as Der Derian's argumentive strategy is in this piece, it can be a little unsettling too and seems, from my own experience using it with students, to elude some readers.

PART 1

Responsibility

CHAPTER 1

Revealing the Hegemonologue

My earliest inkling of a personal connection to colonialism came at the age of seven or eight. Enrolled in the Wolf Cubs of Canada (a junior section of the Boy Scouts), I returned from a weekend outing with a comic book that told the tale of how Robert Baden-Powell came to found the World Scouting Movement. During his service in the Boer War, Baden-Powell had enlisted the assistance of boys to scout in support of the British regulars defending Mafeking. The comic book graphically portrayed the story of these predecessors of the Boy Scouts and how they distinguished themselves in the war. I remember being fascinated by the illustrations that were used to tell the story: images of heroic battles in faraway places, of splendidly uniformed British troops and fierce-looking deep-purple-skinned Zulu warriors. I had a vague sense that my own little Wolf Cub uniform connected me to all of this and, from that day on, I saw our weekly pledge—that included the promise to do our duty to God and the Queen[1]—in a rather different light. Of course, this was neither my first nor my most profound confrontation with the legacies of colonialism. Indeed, as a preschooler, a favorite record was an audio play of the life of American frontier icon Davy Crockett in which he triumphed over the dangers of the "wilderness," battled hostile "Indians," and was sent to Congress—all the elements of an epic story of the righteous conquest and civilization of North America. And, with my friends on our block, more than a few warm summer evenings were spent playing "Cowboys and Indians." But it was the comic book that brought forth my first—albeit naïve—flash of consciousness of the significance of such things and of my own connections to them.

If we do not always recognize our connectedness to colonialism, we are all the less likely to appreciate our complicities in it. In fact, there is something almost counterintuitive about the idea that we might be implicated in colonial practices—after all, decolonization quite famously took place with the final collapse of the European empires in the aftermath of World War II, an era that for many of us, and certainly for our students, has been experienced only vicariously through accounts of the past. And who could know this history better than students of International Relations? Even introductory under-graduate textbooks are quite clear on the matter, pointing to the proliferation of national independence struggles during the period from the 1950s up to the 1970s. But the passing of the era with which we typically identify decol-onization should not be taken to mean that all of Europe's conquests have now been undone. Much to the contrary, vast territories—notably, Australia, New Zealand, and the Americas—remain under the advanced colonial authority of European settler states that displaced or destroyed Indigenous peoples, suppressing their traditional lifeways and forms of sociopolitical organization. Moreover, none of this is properly relegated to the comfortably distant past or ascribed solely to the misguided schemes of our forebears. Rather, advanced colonialism and all of its attendant violences survive not only on the strength of the legacies of earlier times and deeds but, as we shall see in later chapters, are also very much contingent upon ongoing practices. That is to say, both displacement and destruction of Indigenous peoples are ongoing and, in this sense, what we might like (or perhaps need) to think of as the unsavory business of yesteryear is, in fact, still being perfected.

Looking back, the colonial complicities of the comic book, the Davy Crockett record, and games of "Cowboys and Indians" all seem abundantly clear. In its own way, each valorized the violences of colonialism and, more particularly, each was allegorical in the sense that it rendered colonial conquest as natural, inevitable, and desirable. The outcome in our games of "Cowboys and Indians," well scripted by the Westerns we saw on our televisions, was never seriously in question. We were thus recipients and, in turn, performers of the scripts of the dominating society; we were simultaneously repositories and (re)producers of hegemonic narratives of the colonial encounter. Both directly and by extension, we also participated in the (re)production of knowl-edges about Indigenous people(s) that, denying the plausibility of their accounts of themselves and their traditions, can rightly be counted among the violences of colonialism. Importantly, Indigenous voices were absent from these various moments of representation—even when the "Indians" on our televisions (and it was not at all uncommon for them to have been portrayed by Euro-American actors) occasionally uttered more than a war cry, they were still made to speak the scripts of the dominating society. Each a technology for the (re)production of advanced colonial knowledges and, no less, a testament to my own place of privilege as a beneficiary of colonialism, my Wolf Cub uniform, my record, and my role-playing games all underscored my own implication/enlistment in these erasures.

Not all advanced colonial practices are so readily apprehendable, however. Among the less conspicuous are those played out through the production of knowledges seemingly unrelated to Indigenous peoples. The connection is perhaps least discernible in the realms of those social sciences that, like International Relations, have never paid any particular attention to Indigenous peoples. In spite of their apparent remoteness from the machinery of advanced colonialism, however, scholars working in this and other disciplines have been very much involved in the (re)production of its ideational foundations. Indeed, the simple fact of our neglect of Indigenous peoples reflects an enduring deference to one of the most fundamental notions of settler state colonialism: the idea that Indigenous peoples do not constitute authentic political communities. By way of their omissions as much as their claims, scholars of International Relations, like those working in other disciplines, unwittingly participate in the (re)production of the enabling narratives of advanced colonialism. Whether conservative or emancipatory in inclination, the various conceptual treatments of things like security or the good life as well as the broader theoretical approaches to international relations all effect violences through the denial/erasure of Indigenous values and knowledges. This belies any sense that the apparent diversity of theoretical commitments—from the canon of the orthodoxy to the avant-garde of "critical" approaches—precludes their mutual implication in a particular politics. Indeed, it turns out that in some crucial respects they are profoundly monological, speaking in unison the voice of a knowing Western subject that, owing to its universalist pretensions, generalizes and naturalizes concepts, categories, and commitments of the dominating society. And it is thus that they merge with more vulgar—even instrumental—accounts and performances in what might be termed the "hegemonologue" of colonialism/advanced colonialism: that decidedly Western voice that speaks to the exclusion of all others, heard by all and yet, paradoxically, seldom noticed, the knowledges it bears having been widely disseminated as "common senses" rather than as politicized claims about the world and our ways of being in it.

This is not to say that the content of the hegemonologue is not contested, particularly in the academy. On the contrary, it is increasingly fragmented in some of its aspects as disciplinary orthodoxies throughout the social sciences and humanities—and even in the so-called hard sciences[2]—have been confronted with a widening range of critical rejoinders over the last two or three decades. But, as we shall see, even the most radical challenges, unable to shed

Western egoisms and cognitive predispositions entirely, remain implicated in the violences of advanced colonial domination—however trenchant the critiques, it turns out that in some important senses they are critiques from within. There is thus a core constituent of the hegemonologue that seems remarkably resistant to reformulation. And like the monological narrative of colonialism spoken through other technologies of knowledge (re)production—comic books, audio plays, and children's games among them—the hegemonologue is performed in International Relations without intervention by Indigenous voices. This speaks again to the persistence of decidedly European ideas about the field's appropriate objects of study and what counts as legitimate knowledge about them. Similarly, epistemological prejudices deny the validity of Indigenous ways of knowing and forms of evidence, which might include such things as intuition, traditional values, and community consensus (Dyc 1994: 226). And formidable structural-cultural barriers also work to exclude Indigenous voices from direct participation in academic discourse: Lakota social norms, for example, abhor the sort of direct criticism and open disagreement that are central to the production of knowledge through academic debate (Dyc 1994: 226).[3]

The more readily perceptible of the narrative practices of advanced colonialism—popular accounts of the "winning" of the American West, for example—are sustained by its hegemonologue. This story, told by the authoritative knowing and speaking subjectivity of Western academic and political discourse, originates in a privileged European/Euro-American, typically male, voice. But the hegemonologue regularly elides detection. This, again, is because it not only conforms to the common senses of the dominating society, it also underwrites them. We may well be tempted to render blatantly functionalist assessments of the more vulgar knowledge producing/sustaining technologies of colonialism: nineteenth-century pop-culture treatments of Indigenous people(s) that constructed them in ways that seem to exonerate the colonial conquest, for example. It would be rather more problematic, however, to suggest some conscious instrumentality on the part of scholars working in International Relations when they participate in the hegemonologue. Though problematic in all instances, such a view is especially unpersuasive in the cases of scholars who have made it their project to destabilize that "privileged European/Euro-American, typically male, voice" that I am identifying with the hegemonologue. And yet, as we shall see, even explicitly emancipatory projects give ready sustenance to the hegemonologue. Here, then, the complicities with ongoing advanced colonial domination are much more subtle than a straight functionalist account will allow. They are also much more potent.)

Answering the Hegemonologue

In several key respects, postmodern or poststructural strategies of critique are especially well suited to exposing the hegemonologue and interrogating it at its foundations. Informed by the work of Michel Foucault, colonial discourse analysis—and the objects of scrutiny here should include the canons of European social philosophy and their later articulations in, for example, orthodox international theory—is an indispensable means by which to lay bare a host of arbitrary conceptual commitments, such as the ontologized notions about human nature that have enabled the idea of the Primitive. Equally indispensable to a demystification of the hegemonologue is Foucault's notion of "capillary" power. This, according to Foucault, refers to "the point where power reaches into the very grain of individuals, touches their bodies and inserts itself into their actions and attitudes, their discourses, learning processes and everyday lives" (Foucault 1980: 39). Power in this sense is not a finite resource, but inheres in relationships in all areas of social life; it is not directed only against the dominated or oppressed, but "invests them, is transmitted by and through them" (Foucault 1977: 27).[4] Through their acceptance of prevailing definitions of normalcy and/or deviance, for example, people participate in the maintenance of oppression(s).

This is revealing of how the hegemonologue works through us in our scholarship—even in International Relations—and in our everyday lives. The materiality of this is captured in what David Campbell has termed "discursive economy." As Campbell describes it, this is "a managed space" in which those who—enabled by unequal social relations of power—make "investments" in particular accounts of what is real are able to draw "dividends" on them (Campbell 1992: 6). This usefully highlights the materiality of discursive constructions of the colonial encounter borne in comic books, audio plays, and children's role-playing games. And discursive economies are as readily identifiable in the scholarly production of theory as in these more visibly oppressive workings of advanced colonialism: as Arjun Appadurai tells us of Anthropology, for example, "[t]he most resilient images linking places and cultural themes, such as honor-and-shame in the circum-Mediterranean, hierarchy in India, ancestor-worship in China, *compradrazgo* in Hispanic America, and the like, all capture internal realities in terms that serve the discursive needs of general theory in the metropolis" (Appadurai 1988: 46).

Derridean deconstruction also promises to denaturalize many of the central truth claims of the hegemonologue by exposing their indeterminacy. Here, the poststructuralist unsettling of the binary oppositions of Western philosophy is especially valuable. To be sure, there are some dichotomies, such as up/down or left/right,[5] which might approach universality—though not always opposed in radical separation, as the yin-yang formulation of Taoism confirms. But, diacritically constituting Selves over and against essentialized Others, a host of considerably less benign oppositions are bound up in the hegemonologue: rational/irrational, literacy/orality, logos/mythos, civilized/savage, among others. These are not universal—indeed, they are inimical to many Indigenous philosophical traditions, especially in their oppositional rendering. They have, however, been central to the ideational dimension of colonialism/advanced colonialism.

Deeply rooted in Western metaphysics and appealing to its derivative "common senses" about the natural world, binary oppositions are not easily dislodged. This is because masculine/feminine, culture/nature, and the like speak not only to what we think we know about the world but are foundational to our accounts of ourselves as well. Following Jacques Derrida, we can treat binary oppositions as hierarchical[6] inasmuch as the former pole in each of these dichotomous constructions is an ideal figuration of the hegemonic knowing subject, so that each becomes one with a privileged Self. That is to say, more than just sharing a scriptive positionality with the primary markers in various other dichotomies, each is mapped over the others so that, in sum, they are treated as though they were a unified set of attributes. Rational-literate-logos-civilized-and so on thus becomes a unitary formation. Similarly, mapped over with/onto the Other are the polar opposites, each of which is the corruption or negation of the ideal: masculine *versus* feminine, culture *versus* nature. The Other is thus the corruption or negation of the Self—an object of loathing to be variously feared, resisted, or assailed. And as with the Self, essential characteristics of the Other become unified as irrational-orality-mythos-savage-and so on. Ultimately, however, each idea is forever producer and product of its polar opposite, since each is imbued with meaning as much by what it is not as by what it is. This suggests that the knowledges we (re)produce, whether in the academy or in our daily lives, are pregnant with meanings that we might not intend but which we cannot adequately disavow. In short, all knowledge is violent. Derrida proposes that this violence might be interrupted by deconstructing binary oppositions: this involves exposing the binary as a unitary formation, destabilizing the hierarchical separation of the poles by reversing them, and substituting a neologism that encompasses both poles in order to recover their unity (Derrida 1981: 41).[7]

Derrida's notion of *différance*[8] is especially useful in this context. Rejecting the Western philosophical commitment to what he calls logocentrism—the possibility of making truth "present" through an appeal to a point of certain knowledge—Derrida employs Philosophy's own pretensions in order to argue that meaning (re)presented through language necessarily eludes our apprehension. If logocentrism requires a language uncontaminated by external systems of meaning—through the use of metaphor, for example—in which there is an unsullied correspondence between signifier and signified, then it is doomed from the outset since, as he argues, the signified is always constituted through other signifiers (Derrida 1974: 15). That is, our understanding of that which we would (re)present is made possible only by way of reference to other significations, as in binary oppositions. Since this is also the means by which we constitute the signified and hold it up to ourselves in our minds, the problem is not reducible to the imperfect communication of a transcendent truth that we might otherwise apprehend. This means that all signifiers are imperfect (re)presentations of their objects and all bear "traces" of those other signifiers in relation to which it has been possible to think them.[9] Meaning, therefore, is always deferred. Put another way, all (re)presentation is inherently metaphorical (Derrida 1974: 15).

Derrida's coining of the term *différance* is actually a clever word play that is directly demonstrative of the point he wants to make.[10] As a signifier it is ambiguous, its root word, *différer*, expressing the concepts of both difference and deferment. At the same time as it highlights the problems of signification, however, its evocation of difference and deferment connotes, respectively, the spatial and temporal planes across which meaning is dimly plotted. For Derrida, difference and deferment are mutually inherent in the act of signification. *Différance*, then, signals both the process of definition through comparative similarity/difference and the deferment of aspects of meaning that are not immediately present. In their interpolative playing through colonialism, the binaries of Western metaphysics enable a definition of the rational, masculine, civilized Self in opposition to an emotive, feminine, savage Other. And just as opposition constitutes civilized as against savage, so too the ascription of affinities constitutes civilized as rational and masculine while savagery is rendered as emotive and feminine; the binaries map over one another, producing deep matrices of definition through *différance*. At the same time, however, they defer their own deeper meaning as articulations of the very violences of colonial domination that have made it possible to think them in the first place. That is to say, the idea of the civilized Self turns on the idea of the savage Other, which is an idea made possible by the colonial encounter through which savagery was conceived and ascribed. The violences of colonialism/advanced colonialism are therefore a deferred part of the meaning of the Self.

What this tells us is that every text is inherently intertextual, bound up in a complex web of referential and deferred meaning. In its capillary form, power works through the intertext in ways that cannot always be anticipated or controlled by the purposeful machinations of its author-performers. This working of power is mystified, however, inasmuch as a complete reading of the seemingly singular text is confounded by its decidedly unsingular nature. It is thus that Roxanne Lynn Doty finds the practice of foreign policy, which depends upon knowledges constructed both within and without the circuits of decision-making, to be far more expansively sited than might otherwise be imagined (Doty 1993: 303). And if this calls on us to look to less obvious sites when inquiring into the practice(s) of foreign policy, so too a fuller reading of international theory requires that we acknowledge that its texts can never be completely present. We must therefore endeavor to trace its intertextual constitution, peering into less obviously relevant sites of knowledge (re)production with which it is woven together in the hegemonologue. If texts are not hermetic, neither is international theory.

All of this quite strongly recommends an intertextual strategy of reading International Relations—what James Der Derian has described as the attempt "to understand the placement and displacement of theories, how one theory comes to stand above and silence other theories, but also how theory as a knowledge practice has been historically and often arbitrarily separated from 'events,' that is, the materially inspired practices comprising the international society" (Der Derian 1989: 6). This, of course, requires that we resist the systematic marginalization imposed on some theoretical approaches and work to effect the presence of all constituents of the intertext. But it also enjoins us to cast about a bit more broadly than the confines of disciplinary International Relations when we seek to identify "international theory"; it should move us to look for the intertextual workings of international theory in contexts that might seem quite remote from the discipline and its subject matters. Put simply, an intertextual approach should not make the mistake of presupposing—and should at all times be on guard against—arbitrary limits upon the play of intertextuality.

A sensitivity to *différance* and the intertextual relationship between signifiers exposes the inseparability of Self from Other—each bearing traces of its opposite, neither can be rendered discretely. Perhaps the most significant implication of this is that if each pole of a binary opposition derives essential elements of its meaning from its antithesis, then it would seem that neither can precede the other. Whence, then, comes the original idea? Here Derrida's notion of the *coup de force* is instructive, referring to a constitutive moment in which a particular Self and its requisite knowledges are called into being. Though productive, this is also a moment of violent erasure since incompatible knowledges and meanings are necessarily suppressed in deference to a stable definition of the Self. And those that survive must be made to fit into places foreordained by the particular meaning(s) inscribed upon the Self in the *coup de force*: a violence made inevitable by the co-constitution of the idea of the Self in conjunction with a host of other ideas, all of which are interconnected through definitional matrices of meaning hierarchically arranged in the mapping of binary upon binary. Impelled by the universalizing commitments bound up in Western conceptions of knowledge, these violences of erasure and inscription play out temporally as well as spatially. The temporal dimension involves an imperative projection of essential knowledges back onto the past in order to establish through the idea of the Self its own fundamental preconditions. For Derrida, this is exemplified in the American Declaration of Independence that, in asserting its own legitimacy through an appeal to a sovereign people whose claim to sovereignty was founded therein, initiated the very conditions of its own possibility/plausibility; it became the foundation for its own validity, resting upon itself in what Derrida calls a "fabulous retroactivity" (Derrida 1986: 10).[11]

What the concept of the *coup de force* quite usefully highlights is that all history is current history. That is not to say that it is entirely arbitrary or that it is without material expressions and consequences. It is to say, however, that history itself is not material in the sense of some ossified structure unalterably inherited from the past. It is felt and experienced, but as a performative practice that must be reenacted in order to endure. Like all knowledges, it lacks a corporeal form by which it might be present in spite of us. And though it might seem to survive quasi-objectively in the pages of books, even written accounts from the past must be read—a practice—in order to be. It very much matters, then, what is read and what is not, and by whom—these texts, after all, do not address themselves to us unmediated by external signifiers. Moreover, the question of the identity of the reader speaks not only to varied faculties of interpretation, but also to the innumerable sites of relative privilege/marginality that affect how history is practiced. The inheritance of history—to the extent that such an "inheritance" is possible—might thus be likened to a wave: though formed of water, the wave is forever being made anew; accordingly, a wave is not now what it was a moment ago, and the water behind it is relevant to its present constitution only for having enabled its passage through time and space. In this sense, it is an *acting upon* water, and not water itself; a wave, like history, is an action, not a thing. We might thus answer Karl Marx's famous admonition that we make our own history, but not under circumstances of our own choosing (Marx 1963: 15), by saying that, insofar as history is a present practice, there is a sense in which, paradoxically, we also make those conditions that are not of our own choosing by participating in their (re[troactive])production. This begins to unsettle linear expressions of being. Moreover, it means that the foundational knowledges of colonialism, to the extent that we find them being (re)produced in scholarship or in the everyday, are more than mere survivances of some earlier era. Instead, they are present practices of our advanced colonial era.

The violences of the *coup de force* also remind us that the hegemonologue is not the art of the soliloquist. That is, the monological narrative practices of advanced colonialism are not spoken only as if to the Self. Rather, their universalist pretensions—backed up, as we shall see, by the tremendous discursive reach of scholarly voice as much as by the performative practices of pop-culture—insist upon their veracity for the Other as well. This spatial working of erasure and inscription, of course, is entirely consistent with the logics of binary opposition: in order for self-knowledges to have validity, knowledges about the Other must be equally valid. And this underscores not only the contingency of ideas about the Self but the violences that necessarily underwrite those ideas as well. To the extent that it can be shown to reproduce these knowledges, then, scholarly writing is but one point of our insertion into a hegemonologue through which we violently speak the modern Western Self over and against Indigenous peoples. Thus, with specific reference to ethnographic writing, Johannes Fabian has observed that simply writing *about* the Other is an insufficient condition of domination: "the Other must be written *at* (as in 'shot at') with literacy serving as a weapon of subjugation and discipline" (Fabian 1990: 760; emphasis in original).

(Re)presentation of our Others is made all the more problematic by a radical rethinking of intersubjectivity enabled by Derrida's unsettling of logocentrism. The possibility of representing the Other's knowledges without effecting new violences is contingent upon reliable intersubjective understandings in order that those knowledges might be imparted to us in the first place. Intersubjectivity presumes a shared appreciation of meaning between two or more subjects. But if, as Derrida has it, all (re)presentation of meaning—even in the mind—is made possible only through reference to imperfect external signifiers, then metaphor necessarily intrudes even in the moment of our holding up a meaning to ourselves. And the same must also hold for those other subjects with whom we presume to share intersubjective understandings. The idea of intersubjectivity, then, mystifies the heterogeneity of meaning made inevitable by the interventions of multiple subjectivities mutually engaged in signification. Though we might agree upon the application of a single signifier in reference to some idea, the idiosyncrasies of each subjectivity at play will derive some unique meaning from the field of external signifiers. The possibility of a truly shared understanding is therefore put seriously in question. The agreed signifier, however, conceals all of this so that we speak as though of the same understanding.[12]

This problematization of intersubjectivity speaks directly to the limits upon our ability to represent our Others and their knowledges. The very project of seeking to apprehend those knowledges presupposes the possibility of intersubjective understanding—it must be possible for the Other to make them plainly observable to us, the easy metaphor being that they must somehow be brought to light. Like the idea of intersubjectivity, however, this metaphor elicits particular methodological and epistemological commitments. As Peter Hulme and Ludmilla Jordanova point out:

> Light was a central metaphor for knowledge long before the Enlightenment, but at that period it took on new vitality. . . .[T]here was a whole epistemology behind the use of images of "light" in the eighteenth century, one that was boosted by the belief that all knowledge came from the senses and that vision was queen among the senses, with observation at the heart of the acquisition of solid knowledge of the world. Enlightenment was less a state than a process of simultaneous unveiling and observation. To look well and carefully sufficient light is required, and looking in this way was deemed the only route to secure knowledge, although even *a priori* knowledge could be analogised to vision as the product of inner light. (Hulme and Jordanova 1990: 3–4)

But as Derrida has shown us, apprehending the knowledges of the Other is not at all so straightforward an undertaking as the metaphor and its subsumed commitment to empiricism suggest. Rather, signification remains an exercise in the marshalling of external signifiers, de/reconstituting and bending meaning, much as a prism de/reconstitutes and "bends" light. Intersubjectivity, then, can be likened to a gazing upon light refracted through a prism: each of us will experience this differently as the visible light spectrum is disaggregated and uniquely projected from each of its many angled surfaces. This corrects the metaphor by re-rendering intersubjectivity in a way that exposes it as an interplay of subjectivities (an intertextual moment in its own right), resulting in a prismic experience of meaning. And the implications of this become all the more arresting if we imagine having known light only as refracted through the prism.[13]

Profoundly disturbing the ideational foundations of advanced colonialism, postmodernism and poststructuralism lend well to destabilizing grand narratives, decentering the subject, unsettling linear expressions of being, and enabling the possibility of multiple histories. In particular, their critique and rejection of objectivist epistemologies is important inasmuch as the objectivist separation of subject and object sustains some of the worst of the Othering perpetrated through orthodox social theory and, as we shall see in chapter 2, in the disciplinary division of knowledge as well. And whether through colonial discourse analysis or deconstruction, postmodernism and poststructuralism have also proved indispensable to exposing our own colonial/advanced colonial complicities as borne in scholarship and in the everyday.

The Limits of Postmodernism/Poststructuralism: Caveats and Red Herrings

Postmodernism and poststructuralism, of course, are not without problems of their own. Both have drawn considerable criticism in recent years—so much so that it is something of an understatement to call them controversial. Some critiques, though, are more damning than others. Among the less persuasive is the all too frequent objection that postmodernism (the generic label most often applied by the critics) is nothing more than a fashionable "unreason" (Weightman 1998) veiled in an obscure and narcissistic language concealing, at best, naivety, at worst, sophistry. Though possibly the weakest critique of "postmodernism," this view has nevertheless been very much in vogue in recent years, having been brought quite dramatically to the fore in 1996 by way of a hoax perpetrated on the journal *Social Text* by theoretical physicist Alan Sokal. While it speaks also to the viability of the theoretical approach I am adopting, the real value of Sokal's hoax is as an analogy to the sorts of distortions to which, as we shall see in later chapters, Indigenous knowledges are regularly subjected through the impositions of hegemonic cosmological commitments—for this it is worth considering briefly.

Eager to expose what he regarded as postmodern sociologists' misappropriations of concepts and terms from the so-called hard sciences, Sokal composed an article in which he facetiously posited an intersection between postmodern thought and quantum mechanics (Sokal 1996a). When *Social Text* published the article, Sokal revealed his prank and, simultaneously, no small measure of contempt for "postmodern" social theory.[14] The sting lay in his having managed to place in a reputable journal an article that he himself later described as having been "liberally salted with nonsense" (Sokal 1996b: 62). Worse yet, he proposed that his essay had been accepted for publication in spite of its deficiencies because it "sounded good" and "flattered the editors' ideological preconceptions" (Sokal 1996b: 62). Sokal's hoax must indeed have been an embarrassment for the editors of *Social Text*, but there is precious little in his "gotcha" that seems an unanswerable indictment of postmodern or poststructural social theory.

Apparently not content to retire from the offensive, Sokal subsequently teamed with Belgian physicist Jean Bricmont to produce a provocatively titled (*Impostures Intellectuelles*) book-length attack on the work of Jacques Lacan, Julia Kristeva, Luce Irigaray, Bruno Latour, Jean Baudrillard, Gilles Deleuze, Félix Guattari, and Paul Virilio.[15] Together Sokal and Bricmont unleash a fury of objectivist indignation, denouncing these theorists as academic imposters. But they also maintain, unreflexively, some conceptual commitments that preclude their engaging the theories of which they disapprove on their own terms. Instead, Sokal and Bricmont dictate the terms of validity, finding fault whenever those terms are breached. Regrettably, however, the principles they invoke as they sit in judgment are sometimes quite explicitly rejected by the very theorists they would have conform to them—small wonder that the objects of their scorn do not do well in living up to standards of good scholarship dictated by the very things they are trying to problematize.[16] In concluding their indictment of Lacan, for example, Sokal and Bricmont apply the label "secular mysticism" to his writings and those he has inspired, complaining that they are of a genre that addresses itself neither to pure aesthetics nor to reason (Sokal and Bricmont 1997: 39).[17] Here they invoke a binary with categorically opposed poles as the expression of two mutually exclusive literary possibilities—a construction that none of the authors they survey is likely to accept. In light of this persistent failing to read *through* the texts they critique, Sokal and Bricmont seem not to be aware that they frequently make a much better case for the defense than for the prosecution.

A case in point is their treatment of Luce Irigaray, whose critique of phallocentric culture delves deeply into the social construction of sexual difference as a problematic of a project of "sameness."[18] That is, in the representation of the (masculine) Self through the (feminine) Other, the Self is constituted as an ideal type that relegates the Other to a perpetual absence, such that it becomes possible to conceive humanity as "Man." The Self thus becomes a universal referent in an economy of signification wherein the former pole of the masculine/feminine binary is immediately and forever present while the latter is always marked by an absence. Accordingly, Irigaray argues that " '[s]exual difference' is a derivation of the problematics of sameness, it is, now and forever, determined within the project, the projection, the sphere of representation, of the same" (Irigaray 1985b: 26–27). [The Other, rendered only as a distorted form of an ideal, is denigrated in the process—as are all those characteristics associated with the Other through the mapping over of binary upon binary.] Irigaray is careful to avoid any essentializing moves, preserving the sense that in their particularities these are entirely social processes and not the workings of nature.

Unknowingly exposing a prismic working of intersubjectivity, Sokal and Bricmont not only miss this important point but seem entirely unaware of having missed it as well. Instead, their misreading of Irigaray leaves them aghast that she has cast men as rational and objective while characterizing women as emotive and subjective—a move that, they complain, renews some of the worst prejudices of machismo.[19] Had Irigaray done this, their outrage would most certainly be justified, but she has not. Nowhere does she naturalize the binaries that effect the absence of women. Still, this does not stop John Weightman—an admirer of Sokal and Bricmont—from answering Irigaray's critique of science (as a gendered construct implicated in women's oppression) with an absurdity that reflects the same misreading made by Sokal and Bricmont: Irigaray's critique, according to Weightman, "dooms the sexes to eternal intellectual separation" and is as much as "to say that water boils, or should boil, at a different temperature, according to whether the kettle is switched on by the husband or the wife" (Weightman 1998: 480).[20] Such silliness, of course, could only be born of the mistaken impression that Irigaray would seek validity for her claims in correspondence with nature— an error of interpretation that reflects the reader's failure to interrogate Irigaray's text on its own terms. As she is quite clearly referring to arbitrary social constructions, Weightman has missed the mark rather spectacularly. Interestingly, though, he does not seem entirely averse to essentialized oppositions, as betrayed in his closing lament that "even in a highly sophisticated society, there can be a reservoir of unreason waiting to be tapped" (Weightman 1998: 489). Here reason and unreason seem to be mapped with, respectively, sophisticated and unsophisticated societies—in asserting the (present) former, Weightman inscribes and denigrates the (absent) latter, amply illustrating the merit in Irigaray's thesis.

It might be comforting if we could dismiss Sokal and Bricmont as reactionaries whose intervention is founded only in an unthinking rejection of emancipatory politics. We are permitted no such comfort. Their objections are genuine, if sometimes problematic. Indeed, even as he revealed his hoax, Sokal was at pains to point out his political sympathies with the very projects he hoped to discredit: "On nearly all political issues—including many concerning science and technology—I'm on the same side as the *Social Text* editors" (Sokal 1996b: 64). In fact, his original deception seems to have been inspired, at least in part, by a concern that the popularization of "epistemic relativism" on the political Left amounts to a quitting of the field, as it were, leaving purposeful political action as the exclusive preserve of conservatism. Lauding the Left's "worthy heritage" in the political use of "rational thought and the fearless analysis of objective reality (both natural and social)," Sokal goes on to proclaim himself "a leftist (and a feminist) *because* of evidence and logic, not in spite of it" (Sokal 1996b: 64; emphasis in original). That he makes this statement very much in earnest owes to an unflagging confidence in rationalism and objectivism as the essential bases not only of legitimate knowledge but of a viable praxis as well. "Why," he asks, "should the right wing be allowed to monopolize the intellectual high ground?" (Sokal 1996b: 64).[21]

Sokal is by no means alone in reading postmodernism and poststructuralism as ineluctably bent on a nihilistic descent into the abyss of infinite relativism whence nothing can be known and purposeful action is paralyzed. Seemingly, no end of critics from both the Left and the Right have leveled precisely this charge.[22] But such Cartesian anxiety is not at all as well founded as some seem to believe. To be sure, the work of a Foucault or Derrida or Irigaray might be abused into an intense solipsism by the philosophically naïve, but this is a weak indictment of the misused theories themselves. Moreover, it is often the critics, not postmodern or poststructural theorists themselves, who make the leap into the abyss: Sokal, for example, attributes to "postmodernists" the claim that there is no objective reality (Sokal 1996b: 63), thereby misrepresenting the more particular argument that our efforts to apprehend and represent objective reality are never unmediated.[23] Once more, then, Sokal lends support to that which he would discredit, in this instance by betraying the intervention of his own subjectivity in either his grasp of an idea or in the retelling of it, or both. And the distortive intrusions of his subjective commitments do not end there: in his use of the homogenizing label "postmodernism," as though this connoted a unified and undifferentiated body of work, he asserts a totalizing narrative that apparently collapses difference into a sameness of his own making—that is, one in which "postmodernists" deny the existence of an objective reality. It is tempting to call this a neat trick, but Sokal's obvious sincerity reconfirms it as an unintended slip and, therefore, an argument in support of the very ideas he would have us reject. In a wonderfully ironic illustration of Irigaray's thesis, the presence of the "postmodernism" he constructs is contingent upon the absence of her actual argument—and those of Foucault, Derrida, and others, for that matter.

A stronger line has been advanced by critics who are troubled precisely because postmodernism and poststructuralism, contra Sokal, do not seem to them to herald a withdrawal from politics at all. These critiques, which tend to be rooted in a much more careful reading-through the theories they engage, suggest instead that there are important senses in which they (unintentionally) advance a conservative politics directly. In an oft-cited essay, for example, Nancy Hartsock has expressed considerable unease at the historical context in which postmodernism and poststructuralism emerged:

> Somehow it seems highly suspicious that it is at this moment in history, when so many groups are engaged in "nationalisms" which involve redefinitions of the marginalized Others, that doubt arises in the academy about the nature of the "subject," about the possibilities for a general theory which can describe the world, about historical "progress." Why is it, exactly at the moment when so many of us who have been silenced begin to demand the right to name ourselves, to act as subjects rather than objects of history, that just then the concept of subjecthood becomes "problematic?" Just when we are forming our own theories about the world, uncertainty emerges about whether the world can be adequately theorized? Just when we are talking about the changes we want, ideas of progress and the possibility of "meaningfully" organizing human society become suspect? And why is it only now that critiques are made of the will to power inherent in the effort to create theory?   (Hartsock 1987: 196)

These are questions that we ought to take seriously. Importantly, Hartsock does not suggest that the silencing of oppositional politics is an ulterior motive of postmodernists and poststructuralists. On the contrary, she recognizes and applauds the emancipatory hopes underlying much postmodernist and poststructuralist thought.[24] But she also worries that whatever unsettling of the status quo of power relations they might promise comes at the cost of disabling transformative possibilities as well. It is at least noteworthy, then, that the ideas of Derrida, Foucault, Irigaray, and others, whatever those theorists' emancipatory hopes and commitments, were born in the Western academy. Hartsock thus demurs that postmodernists are "the inheritor[s] of the disembodied, transcendent voice of reason" (Hartsock 1987: 200). At very least, it can be said that they do quite often enjoy the inheritance of the seat (or Chair) of privilege whence that voice issued. They might speak from the margins of the scholarly world, but that is often the only sense in which theirs could be called marginal voices.

Raising a similar critique from a postcolonial perspective, Sankaran Krishna notes, with deliberately foregrounded irony, that poststructuralist theory in International Relations sometimes results in readings of the empirical that bear disturbing similarities to hegemonic accounts and narratives (Krishna 1993). Prime examples are the analyses of new weapons technologies and changing ways of warfare undertaken in William Chaloupka's *Knowing Nukes* (1992) and James Der Derian's *Antidiplomacy* (1992). While he finds much of value in these authors' emphases on speed, technology, and new forms of representation as defining features of the 1991 Gulf War, Krishna cautions that this also has the effect of mystifying its material aspects, not least the frightening body count (Krishna 1993: 397–99). Accordingly, he reminds us that "in many ways the Gulf War was very much in the mold of previous conflicts," pointing out that the vast majority of the bombs delivered against Iraq were not precision-guided (Krishna 1993: 398). In emphasizing novel technologies and their effects, then, "postmodernist analyses wind up, unwittingly, echoing the Pentagon and the White House in their claims that this was a 'clean war' with smart bombs that take out only defense installations with minimal 'collateral damage' " (Krishna 1993: 399).[25] And, like Hartsock, Krishna is also concerned about the disabling of an oppositional politics— even as he holds that postmodernism and poststructuralism have "abundantly proven their worth" (Krishna 1993: 396)—that results from an implicit loss of subjectivity.

How, then, do we preserve a viable political praxis? The point of insertion of an oppositional politics may reside in what both David Campbell (1994) and Jim George (1995) have proposed as an "ethics of responsibility."[26] Drawing on the work of Emmanuel Levinas, they offer this as a "first philosophy" that joins a decentering of subjectivity (through an attentiveness to its constitution in relation to the Other) with a refiguration of ethics as a right to be (in relation to the Other) Our responsibility to the Other in this formulation is the founding condition of our own subjectivity, itself possible only as an outcome of a relationship with the Other (Levinas 1986). This produces a much more politically enabling view of ethics than that allowed by the ontologized egoism-anarchy regime of orthodox international theory that, as George observes, fixes boundaries upon the realm of ethical possibilities according to "a rigid power politics logic, which, not surprisingly, gives politico-ethical legitimacy to great power dominance, and, more recently, hegemonic systems of global order" (George 1995: 196). An ethics of responsibility thus liberates political praxis from the austere range of possibilities imposed by this or that totalizing narrative. Moreover, as Campbell points out, it will not permit us the luxury of choosing whether to become politically engaged or not, of declaring that the plight of the Other is none of our affair, since "no self can ever opt out of a relationship with the other" (Campbell 1998: 176).[27]

This is a very important move, one that quite clearly reveals that a commitment to poststructuralism need not preclude a political praxis. But, for the purposes of the present project, what it cannot tell us is what the particulars of a politics ought to be—we still lack a theory. Put another way, the practical content of "responsibility" is as yet unspecified. This might not seem especially problematic on the face of it—in fact, there is much of virtue about it to the extent that it does not presume to legislate, once and for all, a new narrative as though all we have learned about the contingency and indeterminacy of knowledge had suddenly been forgotten. The very same things that would make the assertion of a new narrative problematic, however, threaten to be replicated *ad infinitum* by—somewhat paradoxically—the improvisations of the moment as the terms of the political are articulated idiosyncratically according to commitments held at the level of the individual. The liberally inclined might well view this as an auspicious outcome, submitting the fashioning of a politics to the polyphonic plurality of the hermeneutic circle. Unfortunately, though, this is a circle populated almost exclusively by Western and Western-trained academics: the children of the hegemonologue. Occasionally they do draw upon non-Western concepts and ideas—although Indigenous North American philosophical traditions are among those that have gone almost entirely unacknowledged—but even these are seldom ever heard except through the already authoritative voice of the scholar. Polyphony, then, can be said to be extant only within the relatively tight confines of a particular ethno-cultural milieu. As this also limits the range of political possibilities that can be imagined, for all practical purposes it (re)produces the exclusion of those that remain "outside." Notwithstanding the advices of its poststructural affinities, then, the precise content of an ethics of responsibility is not decided out of a play of all relevant texts; only those produced or reproducible by the privileged parties to the hermeneutic circle can contribute to the fashioning of a praxis. The intertext thus turns out, in an important sense, to be an intratext.[28]

Postcolonialism and the Unsettling of the Hegemonologue

Besides some clear articulation of a politics, still lacking is an account of the present historical conjuncture. Before engaging directly with issues of Indigeneity, then, what is needed is to specify a theory of (current) history.[29] What are the defining features of our sociopolitical world? What are their antecedents and what has impelled them to unfold in the particular ways they have? And what, if anything, does all of this suggest about transformative possibilities and the fashioning of an oppositional politics? These are questions for which postmodernism and poststructuralism alone cannot and should not be expected to propose answers. As strategies of critique, they do—quite indispensably—show us how we might lay bare and demystify the hegemonologue, the ways in which power works through it (and us), and our own involvements in and complicities with it. To be sure, these are important political moves in themselves—ones that are quite clearly oppositional. What is less clear, however, is the extent to which they are enabling of projects that are more proactively emancipatory in the sense of proposing the terms of a more just sociopolitical order. This is partly in consequence of their leveling effect[30]—a function that calls critical attention to the relative ethno-cultural homogeneity of the hermeneutic circle. But more particularly, the lack of an account of the present historical moment leaves the cast of characters, as it were, as yet unspecified: who is to be emancipated and from whom/what? Put another way, which meanings and whose definitions will we privilege and why?

The assertion of an adequate political praxis and theory of (current) history is complicated by an inherent Eurocentrism that infects both post-modernism and poststructuralism, threatening to subvert their best designs. Poststructuralism (a moment in European social philosophy) follows and is contingent upon the legacies of structuralism (another moment in European social philosophy).[31] And to the extent that modernism is Europe's child, it is problematic to situate other parts of the world in what is sometimes termed a "postmodern condition." To say that we live in a "postmodern era" or are conditioned by a "postmodern attitude" is to define our historical moment against a decidedly European referent.[32] Modernity might have been widely disseminated, but only imperfectly so—it did not take hold everywhere, and even where it did it was subject to the interventions of local subjectivities. Worse yet, the invocation of a "post-" in a world wherein, we might credibly argue, modernism has never been ubiquitous portends recourse to a "pre-" in reference to some contexts. And this treads dangerously close to a reinvigorated conception of linear, though uneven, "progress"—a core component of colonial ideologies from the so-called Age of Discovery, through nineteenth-century evolutionism, and enduring in contemporary liberalism.

The advices and strategies of postmodernism and poststructuralism nevertheless remain indispensable to the present project. But something more is clearly needed before we can draw on them in exploring the intersections between Indigeneity and international theory. Here I turn to postcolonial theory, reading it as a theory of (current) history so as to give political content to an ethics of responsibility. What, precisely, is meant by those who invoke postcolonialism? This is indeed a vexed question—meaning, of course, that there is no firm consensus on an answer. Even a minimal perusal of this expansive theoretical terrain is quite beyond the bounds of this book, so I confine myself to specifying my own commitments, highlighting important caveats as necessary. Apart from this, I do not venture to summarize the wide range of postcolonialisms that have been proposed by various theorists; nor do I try to recount all of the lively debates between them.[33] The reader is therefore cautioned that the "postcolonialism" articulated here is my own preferred version and not at all a settled rendition.

At its most general, postcolonialism is a body of theory that self-consciously engages unequal power relations between peoples by critically interrogating the cultural outcomes of those relationships. Enjoying its greatest influence in the fields of Literary Criticism and, later, Cultural Studies, a primary focus of postcolonial theory has been on the cultural products and practices of the colonized with a view to uncovering the ways in which they have been shaped by the colonial encounter, the crucible being located on the theme of resistance. "Postcolonial" characterizes the literatures written by the Other in the colonial encounter: those that, together with literature written in Britain, used to be called "Commonwealth literature." Postcolonial literature, according to Elleke Boehmer,

> is that which critically scrutinizes the colonial relationship. It is writing that sets out in one way or another to resist colonialist perspectives. As well as a change in power, decolonization demanded symbolic overhaul, a reshaping of dominant meanings. Postcolonial literature formed part of that process of overhaul. To give expression to colonized experience, post-colonial writers sought to undercut thematically and formally the discourses which supported colonization—the myths of power, the race classifica-tions, the imagery of subordination. Postcolonial literature, therefore, is deeply marked by experiences of cultural exclusion and division under empire. (Boehmer 1995: 3)

In its early usages, the term "postcolonial" thus connoted an important essence of the English-language literatures written in the former colonies and influenced by the experience of colonialism and decolonization. From these narrower origins, postcolonial theory has emerged, spawning deconstructive strategies of critique that seek to recover the subjectivity and voice of the colonized.

As part of this broader project, colonial discourse analysis is employed in the reading of Europe's cultural products—principally fiction but including popular historiography and other forms of pop-culture representation and performance as well—through the experience of colonialism in an effort to reveal their complicities with colonial domination. Deeply indebted to Edward Said's work on Orientalism (Said 1979), postcolonialism treats the colonial encounter as a central determinant of the material and ideational culture of the colonizer as well, reading the ideologies of colonialism as articulated—variously implicitly and explicitly—through art, literature, and science.[34] This, in turn, enables the view that Europe was itself forever changed by its own colonial expansion, and in ways that might not always be well appreciated. Postcolonialism is thus invested with at least the rudiments of a theory of (current) history, defining our present moment in terms of the ongoing performance of (advanced) colonial knowledges and practices. And "postcoloniality," as characterized by Boehmer, is its complementary oppositional political stance "in which colonized peoples seek to take their place, forcibly or otherwise, as historical subjects" (Boehmer 1995: 3). For my purposes herein, then, the hegemonologue is revealed not only as the central object of inquiry but also as the focal point of an oppositional politics that would seek to interrupt it by challenging its narratives and working toward the audibility of the marginalized voices it violently erases.

My approach is postcolonial in the sense that it explicitly seeks to destabilize the hegemonologue, to reapportion the authority to narrate, and thus to hear voices marginalized in the colonial encounter, taking heed of the subjugated knowledges they bear. Recourse to postcolonialism might, at first, seem out of place in reference to peoples who, like Indigenous North Americans, have not yet been emancipated from advanced colonial domination. It is important to note in this context that independence and decolonization are not the same thing: that the advanced colonial states of the Americas have, to varying degrees of perfection, won their independence is not to say that the territories over which they now exercise autonomous dominion have been decolonized, the founding mythologies embodied in the American Declaration of Independence notwithstanding. Reminding us that "Native American people today continue to live under an ongoing domestic imperialism," Arnold Krupat argues that contemporary Native American literature "is not and cannot now be considered a postcolonial literature for the simple reason that there is not yet a 'post-' to its colonial status" (Krupat 1994: 169). More generally, Anne McClintock warns that, "in its premature celebration of the pastness of colonialism, ['post-colonialism'] runs the risk of obscuring the continuities and discontinuities of colonial and imperial power" (McClintock 1994: 294).

But postcolonialism is not a synonym for decolonization or independence; it is neither an explicit nor implicit claim that the era of colonialism has passed. Following Homi Bhabha, the idea of a "postcolonial condition" connotes the persistence of the effects of colonialism—articulated through enduring structures of domination and exploitation—that persist even after decolonization (Bhabha 1994: 6). The "post" in postcolonial, then, should not be taken to mean that colonialism is over, even where decolonization has occurred. On the contrary, with the indelible marks it has left on peoples and on histories it endures genealogically in the possibilities it has created and in those it has foreclosed. If the "post" in postcolonial signals an "after," it is in reference to the *experience* of colonialism—and those enduring changes that it has wrought—much more than colonialism itself. Read this way, the post-colonial is as relevant to contexts of continuing direct colonial domination as to the post-independence state; both contexts have been inscribed by colonialism and, though this experience does not define them by itself, neither is it at all insignificant. As Ruth Frankenberg and Lata Mani put it, the "post" "mark[s] spaces of ongoing contestation enabled by decolonization struggles both globally and locally" (Frankenberg and Mani 1993: 294). According to Stuart Hall:

> "After" means in the moment which follows that moment (the colonial) in which the colonial relation was dominant. It does not mean . . . that what we have called the "after-effects" of colonial rule have somehow been suspended. It certainly does *not* mean that we have passed from a regime of power-knowledge into some powerless and conflict-free time zone.   (Hall 1996: 254; emphasis in original)

In short, the idea of a "postcolonial condition" admits readily of the persistence of advanced colonial domination; likewise, postcolonial theory is exceptionally well suited to the present inquiry into the violences visited upon Indigenous peoples by international theory.

Postcolonial theory goes to the heart of the issue that makes it appropriate for me to suggest important intersections between two such seemingly disparate realms as international theory and pop-culture treatments of Indigenous people(s): the power to represent. As Phillip Darby and A.J. Paolini put it, "[a]kin to Marx's owners of the means of production, postcolonialism locates power in the control over the means of representation" (Darby and Paolini 1994: 387). Consistent with the Foucauldian view that power is not a finite resource, however, we should not expect it to be "held" only in the dominating society. That is to say, though there will most assuredly be imbalances, there will never be a monopoly on power. Said has been critiqued on these grounds—specifically, his simple East versus West dichotomy and preponderant emphasis on deconstructing the texts of the West occludes resistance and seems to leave little room for the possibility of transformative change (Loomba 1998: 49). It is worth noting, though, that when we talk about Indigenous North Americans we are also talking about a highly perfected system of domination such as never existed in most of the locales subjected to European colonialism. Here colonial discourses work virtually without audible interruption, so that Said's omissions seem somewhat less problematic.

Still, his often homogenizing treatment of the West is considerably more troubling inasmuch as it makes the possibility of endogenous resistances to the hegemonologue seem remote at best. As Ania Loomba points out, however, Bhabha's view that the binary oppositions of colonialism are never stable—this owing to the impossibility of establishing and securing unmediated meaning—is a useful corrective to Said's emphasis on domination (Loomba 1998: 232). This reclaims the possibility of resistance and change, locating it in the never finished and always shifting negotiation of identity and difference. And it also underscores the ease with which social theory can, through the hegemonologue, become implicated in a conservative politics: to the extent that a given theory (re)produces and naturalizes the binary oppositions and other discursive formations that underwrite advanced colonialism, it mitigates against their instability. This means that even emancipatory theory is disenabling of resistance when it does not eschew these knowledges.

An important theme in much postcolonial literary theory turns on the overriding authority of voices speaking from the centre vis-à-vis those of the periphery. This is not only a matter of domination, of the forcible suppression of the voice of the Other. More than this, it calls upon us to look to the more passive ways in which that authoritative knowing voice originating in the West contains marginal narratives by self-referentially setting the terms of legitimacy. As Bhabha argues, the assessed worth of a text is inversely proportional to the degree of its deviation from the hegemonic narratives and ideas of the dominating society (Bhabha 1984). In a manner that the reader will recall from Sokal and Bricmont's (mis)treatment of postmodernism and poststructuralism, the texts of the West and the knowledges they bear have thus become the standards against which the relative authority of voices is measured. The result is that voices speaking from the margin become audible only when they mimic those of the centre—when they retain too much that is "foreign" or arcane to the Western ear they are typically reduced to artifacts of the exotic, charming perhaps but hardly authoritative. And just as Sokal and Bricmont's failure to read through the texts they critique is a cause of serious distortions when they come to represent elements of them, representations of the texts of the Other too easily suffer the prismic distortions of intersubjectivity, made all the worse by the violent impositions of the hegemonologue. Of course, all of this profoundly limits the extent to which it is possible to articulate a unique perspective (Ashcroft et al. 1989: 6).

Further complicating matters is Gayatri Chakravorty Spivak's famous pronouncement that the subaltern cannot speak (Spivak 1988a). It is not that the oppressed Other is voiceless. On the contrary, Europe's Others do speak out, sometimes vociferously. But the monolithic category of "the subaltern"—like "the Other"—collapses myriad Other subjectivities into a single essentialized voice that elides its own heterogeneity and therefore cannot be heard. Strongly influenced by Derrida, Spivak finds that the voice of the subaltern is lost in a complex play of *différance* that suppresses its own marginal voices.[35] And this prompts her to question any project aimed at constructing a speaking position for the subaltern inasmuch as this can only be premised on an essentialized fiction that suppresses the margins of margin. Thus, if the subaltern cannot speak, it is in the sense that the constructed subject that would speak its voice is constituted through an assumption of homogeneity that does not stand up to scrutiny. Worse yet, this means that even sympathetically conceived attempts at constructing a voice for the subaltern will repeat the representational violences of colonialism. However well intentioned, then, our efforts are fundamentally misguided when we presume to speak for our Others instead of working toward the audibility of their own voices.

There is a powerful warning in this that ought to be heeded by the post-colonial intellectual: when we venture to speak for Others, even with the most nobly conceived emancipatory agenda at the fore, we unavoidably commit acts of violence. What Spivak underscores for us is that ours can never be a substitute for the voice of the Other. Put another way, we must recognize that the hermeneutic circle is not a conversational space—as presently constituted, it is an almost hermetically closed Western discursive space. The voice of the Other is not present and so the Other's knowledges and narratives can only be "brought in" through the mechanism of (re)pres-entation, with all of the problems and pathologies that necessarily entails. True conversation, then, should be our goal. In the first instance this means working toward the audibility of the voices of Indigenous people(s)—among Others—in theorizing the international; it does not mean presuming our-selves unproblematic surrogates in their place. It is not enough to say simply that we will converse as this makes the common liberal mistake of assuming that procedural entitlements are one with the ability to enjoy them. Rather, conversation must be our first methodological commitment in theorizing the international, something actively promoted not passively declared. This calls for a deconstruction of our own privileged speaking positions, which have been secured through the enforced absence of Other voices. And this, in turn, enables a second move in which the authority of the Other's voice is affirmed. To the extent that genuine conversation is established in this way, the hermeneutic circle can be rehabilitated.

A commitment to conversation presupposes a heterogeneous interaction of voices, all possessed of the authority to speak. But postcolonial theory has been criticized on the grounds that it flattens specificity, reducing everything to an expression of colonial power relations. Spivak, for example, is skeptical about the possibility of recovering unaltered precolonial lifeways and identities, seeing the colonial experience as (re)definitive for the peoples it touched (Spivak 1988a). That is not to say that colonial authority remakes Others in ways entirely of its own choosing. Rather, in the colonial encounter, domination and resistance work upon one another—in ways overt and unseen—to change colonizer and colonized alike. "Thus," according to Loomba, "Gandhi's notion of non-violence was forged by reading Emerson, Thoreau and Tolstoy, even though his vision of an ideal society evoked a specifically Hindu vision of 'Ram Rajya' or the legendary reign of Lord Rama" (Loomba 1998: 174). For Bhabha, the "location of culture" is in the liminality repudiated by the sharp either/or segregations of oppositional dichotomies. Here, at the interstices between Self and Other, certain knowledge fades into ambivalence as the colonized make their own subjective interventions in (re)presenting the texts of the colonizer.[36] Although the center speaks its hegemonic knowledges over and against the periphery, its inability to secure the stability of meaning requires that they be constantly in flux. Therefore, in Bhabha's view, mimicry of the colonizer by the colonized is not renderable simply as an expression of solicitous desire,[37] it is also bound up with resistance in its subversion of hegemonic narratives and ideas.

Spivak's skepticism about the possibility of recovering precolonial subjectivities enjoins us to recognize that the workings of colonialism and patriarchy insinuate themselves into these re-renderings in ways that ensure the voicelessness of the subaltern. Essentialized oppositional identities thus portend violent erasures of their own, and this means that all such constructions must be viewed as "strategic" moves,[38] constructed for purposes of collective political action or analytical expediency but not as ontologically enduring identity claims. These are separate from and against theory—not an exegesis, but a license that we grant to ourselves to enable the formulation of an oppositional subjectivity for the purposes of political praxis.[39] For his part, Bhabha's conceptions of mimicry and hybridity find the locus of a politics in the interstitial space that, by the very logic of binary oppositions, must exist but is never foregrounded by them. This destabilizes the enforced absence of the Other who is made present at the ideational equivalent of what Mary Louise Pratt calls the "contact zone" (Pratt 1991).[40]

Spivak's suspicion of the "traditional" and Bhabha's workings of mimicry and hybridity thus bear on how we think about political praxis and emancipatory change. But, each in its own way, they can also be read to imply that all that is meaningful about the postcolonial subject is constituted solely on the basis of the colonial encounter; taking this to its extreme, the subjectivity of the Native might thus be rendered as a passive skin that had awaited the inscriptions of Europe before it could find its own meaning. For Thomas King, this is yet another violent (re)inscription in its own right:

> While post-colonialism purports to be a method by which we can begin to look at those literatures which are formed out of the struggle of the oppressed against the oppressor, the colonized and the colonizer, the term itself assumes that the starting point for that discussion is the advent of Europeans in North America. At the same time, the term organizes the literature progressively suggesting that there is both progress and improvement. No less distressing, it also assumes that the struggle between guardian and ward is the catalyst for contemporary Native literature, providing those of us who write with method and topic. And, worst of all, the idea of post-colonial writing effectively cuts us off from our traditions, traditions that were in place before colonialism ever became a question, traditions which have come down to us through our cultures in spite of colonization, and it supposes that contemporary Native writing is largely a construct of oppression.   (King 1990: 11–12)

Like postmodernism and poststructuralism, postcolonialism seems susceptible to succumbing to a Eurocentrism of its own, violently remaking Others according to a European referent. As King compellingly puts it: "I cannot let post-colonial stand—particularly as a term—for, at its heart, it is an act of imagination and an act of imperialism that demands that I imagine myself as something I did not choose to be, as something I would not choose to become" (King 1990: 16).

While King's objections stand as important warnings about how postcolonial theory might be made to effect erasures, this would require either a misreading or an abuse of the theory itself. The idea of a "postcolonial condition," presuming only a mutual—and, crucially, varied—insertion into an historical process, portends no necessary homogenizing moves. The colonial experience is not treated as though it is by itself definitive of the postcolonial subject, meaning that specificity is still quite easily foregrounded. And this sensitivity to alterity underscores that postcolonialism actually resists homogenization— "the postcolonial subject" is a shorthand that is in no way intended to suggest a uniform experience of colonialism giving rise to an essential postcolonial consciousness or identity. And to say that colonialism is the defining feature of postcoloniality is not to claim that it is also *the* defining feature of postcolonial subjectivities. Of course, its importance is not to be understated either. As Bhabha tells us, "[p]ostcoloniality . . . is a salutary reminder of the persistent 'neo-colonial' relations within the 'new' world order and the multi-national division of labour," and "[s]uch a perspective enables the authentication of histories of exploitation and the evolution of strategies of resistance" (Bhabha 1994: 6). But, for Bhabha, this is not a denial of heterogeneity. Much to the contrary,

> postcolonial critique bears witness to those countries and communities—in the North and the South, urban and rural—constituted, if I may coin a phrase, "otherwise than modernity." Such cultures of a postcolonial *contra-modernity* may be contingent to modernity, discontinuous or in contention with it, resistant to its oppressive, assimilationist technologies; but they also deploy the cultural hybridity of their borderline conditions to "translate," and therefore reinscribe, the social imaginary of both metropolis and modernity.   (Bhabha 1994: 6)

This affirms not only the existence of subjectivities constituted "otherwise than modernity," but emphasizes also that even those who have been touched by modernity are not simply remade in its image and, crucially, that they change what has touched and changed them. Far from denying the persistent influences of precolonial subjectivities, Bhabha insists upon their enduring importance as the basis for Others' agencies and abilities to resist.

Spivak does not rule out a role for "tradition" either. Since strategic essentialism is not a theoretical move—in fact, it is a necessary departure from theory, consciously undertaken—it does not stand apart from any of the Derridean underpinnings of her work in the sense that a constructed oppositional subjectivity could be understood to reside beyond the reach of deconstruction. Every instance of strategic essentialism therefore always calls out to be contextualized and historicized itself. Colonialism is a part of that historicization, but so are local specificities. Spivak's misgivings about precolonial identities, then, need not deny the possibility of substantial survivances; she tells us only that a "pure tradition" cannot be expected to survive in all its aspects, entirely without supplement or alteration. Moreover, not all change is decided for the colonized by the colonizer: the meeting of worlds effects change by way of purposeful adoption as much as through violent imposition. For instance, though a staunch traditionalist, famed nineteenth-century Hunkpapa Lakota leader Sitting Bull did not imagine (nor does he seem to have desired) that all European influences could be refused: "When you find anything good in the white man's road, pick it up but when you find something bad . . . leave it alone" (quoted in Rice 1991: 29–30).

Homogenization has also been charged in the context of the local itself: even if we are clear that "postcolonial subjectivities" are varied with the specificities of multiple locals, there remains a danger that the heterogeneity of the singular local context might be flattened. Arun P. Mukherjee argues that "[w]hen post-colonial theory constructs its centre-periphery discourse, it also obliterates the fact that the post-colonial societies also have their own internal centres and peripheries, their own dominants and marginals" (Mukherjee 1990: 6). This too is a useful caution, particularly when we engage directly with the empirical. But, again, there is nothing inherent in postcolonialism that necessitates the denial of local heterogeneity. Rather, this seems more the sin of nationalism and, in this regard, is quite effectively addressed by Spivak's critique of the Subaltern Studies Group, her notion of strategic essentialism furnishing the corrective.

The influence of poststructuralism in much of postcolonial theory—exemplified in the works of Bhabha or Spivak—has also raised concerns, reminiscent of Hartsock's critique of postmodernism. Once again, the privileged location of the theorists is very much at issue. Kwame Anthony Appiah situates postcolonial intellectuals themselves in a liminal space where they "mediate the trade in cultural commodities of world capitalism at the periphery" as they present the Other to the West and vice versa. For Appiah, then, "[p]ostcoloniality is the condition of what we might ungenerously call a *comprador* intelligentsia" (Appiah 1991: 348; emphasis in original). Phillip Darby has similar concerns:

> The enunciation of the margin as the site of creative thinking at first sight appears a salutary corrective, and an empowering one. But when we listen to the voices associated with the margin and reflect on the messages conveyed there is less cause for confidence. Overwhelmingly the voices are those of people now situated within Western universities—Said, Spivak, Bhabha, Appiah, Mbembe, Mazrui and the like—speaking the language of the contemporary academy.   (Darby 1997: 21–22)

Once more, we are cautioned that we might only be listening to the margins of the *scholarly* world, which itself is far from marginal in a broader sense. Still absent is the voice of the Other.

Considerable concern has also been expressed over the danger that postmodernism or poststructuralism might dilute the potential of resistances enabled by postcolonialism. Linda Hutcheon, for example, argues that postcolonialism has more in common with feminism than with postmodernism: "The post-colonial, like the feminist, is a dismantling but also constructive political enterprise insofar as it implies a theory of agency and social change that the postmodern deconstructive impulse lacks" (Hutcheon 1989: 171). In a similar spirit, Darby is emphatic about the need to move beyond deconstruction:

> By now we have been told *ad infinitum* that everything is constructed. What is required is "peopling" discourses about the Third World and providing footholds for action. There is a demonstrable need to go beyond invocations about the contingent and the indeterminate and to face up to the enormity of the problems.   (Darby 1997: 17)

Darby is quite right, the persistent incompleteness of the hermeneutic circle suggests that we do need to supplement the deconstructive with some sort of goad toward transformative change. And yet, at the same time, the hegemonologue endures—a certain sign that the project of deconstruction has not outlived its usefulness. "*We*" might have "been told *ad infinitum* that everything is constructed," but it is well that we remember too the earlier concerns about who "we" are: a privileged intelligentsia on the margins of the scholarly world. Outside the confines of this small realm, deconstruction remains an engaged and often novel oppositional project.

How, then, do we reconcile the apparently contradictory demands for sustained deconstruction and an oppositional politics? Spivak's notion of strategic essentialism is one answer. Of course, the privileged position of most postcolonial intellectuals gives cause for concern about the kind of oppositional subjectivity that might be constructed and whose voice it will speak. Here the indispensability of conversation is brought into sharpest relief, as is Spivak's joining of strategic essentialism with "a scrupulously visible political interest" (Spivak 1988b: 205). Still, Darby is critical of strategic essentialism, casting it as a "sleight of hand" intended to "establish a basis for thinking positively and politically in the crevices and recesses between the forbiddens of postmodern thought" (Darby 1997: 17). To be sure, there is something of a contradiction at work here, but if postcolonial theory has taught us anything it is that every aspect of life involves the negotiation of contradictions. We should hardly expect the nexus between theory and praxis to be exempt from this. Darby's more particular concern is that, being made to fit in this way "enscribes the secondary status of the Third World in the thinking of the First" (Darby 1997: 17). On this point I am in full agreement, although it seems to me that the appropriate response is to plump for conversation as I have described it earlier.

Although I am receptive to some of the concerns he raises, on the whole I am more optimistic than Darby about the viability of the poststructural-postcolonial synthesis. If poststructuralism *enables* the idea of multiple histories, postcolonialism actively *promotes* it. Rejecting the more vulgar caricatures of poststructuralism, I find the promise of a political project in the idea of an ethics of responsibility—an idea that ought to put to rest once and for all the Cartesian anxiety that has inspired so much alarm. My methodological commitments and my strategies for exposing and unsettling the hegemonologue are inspired by poststructuralism, but my broader theoretical framework is postcolonial. An ethics of responsibility is the coupling mechanism: it is the space for a politics whose content is, for me, given by postcolonialism. And though my own privilege should rightly make such negotiations suspect, I am not persuaded that it would be any less problematic if my theoretical commitments were of a more explicitly programmatic revolutionary nature. Would I then be any less mired in contradiction? Alternatively, could I escape it by withdrawing into the status quo in silence? My sense of it is that I could not.

Importantly, postcolonialism is concerned with the ways in which the experience of colonization affected the colonizers as well. Unless it is imagined that there is something inherent, something essential, in Europeans that makes them naturally dominant, we must recognize that we too have been made by the experience of colonialism. My Davy Crockett record, games of "Cowboys and Indians," and the revelations of the comic book betray my own condition as a postcolonial subject. There is thus a sense in which I share Thomas King's resentment at having been marked as something I did not choose to be and would not choose to become. This is the violence of inscription, a process that, almost by definition, is not a way we make ourselves. And it is not at all unusual to find that it defies our own attempts to name ourselves. But I also hasten to add that it is neither postcolonial theory nor its expositors that have inscribed us. Rather, those names we did not choose for ourselves are inherited legacies of colonialism (re)instantiated and sustained in the hegemonologue of advanced colonialism.

None of this, of course, is to say that I have experienced colonialism/advanced colonialism in ways analogous to the experiences of those who are more readily recognizable as its Others. Nor is it intended to suggest that I am not also a material beneficiary of colonialism. Instead, this very ambivalence only underscores further that who, what, and where I am are all inseparable from colonialism. Since there can be no single monolithic postcolonial subjectivity, to recognize the colonizers—or those who have benefited from colonialism—as postcolonial subjects is not a move that flattens the diversity (and profound inequality) of our various experiences. That is, it does not mystify sites of margin and privilege or deny unequal power relations. On the contrary, it reveals hidden workings of the hegemonologue, resisting the collapsing of heterogeneity into the simple binary opposition of oppressor and oppressed. As Ashis Nandy argues, the "White Sahib," whose own humanity is a casualty of the master-slave relationship, is not an essential tyrant: "Certainly he turns out to be . . . not the conspiratorial dedicated oppressor that he is made out to be, but a self-destructive co-victim with a reified lifestyle and a parochial culture, caught in the hinges of history he swears by" (Nandy 1983: xv). As Derrida might put it, the violences visited upon the slave are a deferred part of the meaning of being the master. Thus, in Nandy's formulation, the apparent "victors" under advanced colonialism "are ultimately shown to be camouflaged victims, at an advanced stage of psychosocial decay" (Nandy 1983: xvi). The significance of this for Nandy is that "[a]ll theories of salvation which fail to understand this degradation of the colonizer are theories which indirectly admit the superiority of the oppressors and collaborate with them" (Nandy 1983: xv).

It is in this sense that ours is a postcolonial world. The European conquest of the planet has left few, if any, places and peoples untouched; it is our common experience notwithstanding the wide divergence in the circumstances of our insertions into it.[41] Accordingly, a central commitment of this book is to the proposition that International Relations is simultaneously subject and object of advanced colonialism. A postcolonial subject in its own right, international theory has been both shaped and constrained by the experiences of colonialism/advanced colonialism. As Sandra Harding reminds us, "[m]odern sciences are 'local knowledge systems' no less than are the science and technology systems of other cultures" (Harding 1998: 178). The hegemonologue works through both critical and orthodox international theory in ways that reveal it as local knowledge. And this is no less true of theory produced in other disciplines. Our potential in theorizing the "international" is limited by the pretensions of a "science"—pretensions that work to frustrate conversation by denying the legitimacy of other bases of knowledge. With the aim of unsettling this very conceit, Paul Feyerabend argues that,

> science is much closer to myth than a scientific philosophy is prepared to admit. It is one of the many forms of thought that have been developed by man, and not necessarily the best. It is conspicuous, noisy, and impudent, but it is inherently superior only for those who have already decided in favour of a certain ideology, or who have accepted it without ever having examined its advantages and its limits.   (Feyerabend 1975: 295)

International Relations encloses a body of local knowledge with nothing more than ill-conceived pretensions to universality. It is in this sense that international theory limits our ability to theorize the international. And this is also the enabling condition of its/our advanced colonial complicities.

Uncovering the Cosmological Imperialism of International Theory

All international theory is cosmologically inflected. The binaries of Western metaphysics are not merely convenient devices drawn upon instrumentally. Rather, they reflect prevailing cosmological commitments whose antecedents can be traced from Greek antiquity, through Biblical scriptures, and into modernity. More than just handy expedients, they are expressions of our most basic and uninterrogated ways of apprehending the world and our experiences in it. But the predisposition to conceive of the cosmos in terms of binary oppositions, although widely disseminated with the spread of the dominating knowledge system, is neither natural nor universal. Rather, the deeply entrenched sense that immutable binary oppositions are ubiquitous in nature owes in large part to another Western cosmological predisposition: linear expressions of both process and being. By way of contrast, the ascendancy of the circle over linear expressions of existence in traditional Lakota cosmology does not lend well to the construction of dichotomies. The unyieldingly linear conceptions of being and rigid binary oppositions of Western cosmology, however, form the core content of the hegemonologue that, in turn, insinuates itself through both orthodox and critical international theory such that one might as readily as the other be implicated in the violences of advanced colonialism.

What is perhaps most troubling about this thesis is that it would seem to render implausible the possibility of any *broadly* emancipatory project drawn from the existing corpus of theory in International Relations. This urges forth the revelation that the idea of counter-hegemony has been framed within the confines of a decidedly Western cosmology that lies beyond the reach of critical inquiry. Western ideas about the universality of knowledge necessarily negate the possibility of multiple cosmologies so that the operant cosmology of advanced colonialism excludes all others. Cosmological ascendancy, then, is an unseen first order of hegemony. A second order, and one that is more familiar, is the dominance of a particular theoretical approach within a discipline. This expression of hegemony is well known to critically inclined scholars who often connect it to a third order: "real world" manifestations of supremacy, as in the unequal relationships between states or peoples.[42] The point here is not to suggest that discrete forms of hegemony operate in separate spheres, but is to enable the observation that counter-hegemony is bounded by context—what is counter-hegemonic in a disciplinary sense, for example, might nevertheless be complicit with the hegemony of the operant cosmology.

As a theory of the universe, cosmology specifies what is and what is not, what can be and what cannot. Cosmology, however, is seldom acknowledged and rarely subjected to scrutiny. Instead, it operates largely beyond our acuity, dictating an account of the natural order, an account of all that is or could be. But it is also a limiting account that does not admit of a fuller range of possibilities. If, for example, a Lakota person were to tell us that the buffalo are her relatives, our cosmological commitments would tell us that this could not be so in any literal sense. As we shall see, however, there is nothing of metaphor or hyperbole about this claim, though what we think we know about the natural order prevents us from seeing the sense in which it is true. This is because Western cosmological ascendancy begets particular limits on ontological and epistemological possibilities and these limits pronounce upon the boundary between the ridiculous and the sublime. This boundary, in turn, safeguards the hegemony of the dominating knowledge system and reinforces cosmological hegemony. A range of ontological and epistemological commitments may nevertheless flow from a given cosmology so that it need not foreordain any particular ontology or epistemology; it can, however, foreclose certain possibilities. Thus, there can be considerable diversity in both these senses among contending theories in International Relations, though all still exclude possibilities deriving from exogenous cosmologies. What cosmology dictates, then, is the field of ontological and epistemological possibilities. Contestation in International Relations takes place only on the terrain occupied by questions of ontology and epistemology and the range of such questions is determined in advance by cosmological commitments that insinuate themselves from beyond the reach of debate. This has the effect of shielding Western cosmology from serious scrutiny without impeding the powerful influence it exerts over knowledge.

The cosmologically sustained Western philosophical predilection to conceive of the universe in linear and binary terms is expressed also in the mode of knowledge production enshrined in the academy. Here different theoretical approaches end up pitted against one another in an uncompromising competition for primacy—the object of scholarly debate, after all, is to persuade our detractors of the errors of their perspectives and of the merits of our own. Accordingly, critical theoretical approaches have been predisposed to manifest not as counter-hegemonic forces but, rather, as *contenders* for "second-order" hegemony. Dichotomous thinking gives rise to "either/or" propositions about truth and knowledge that implicitly cast each approach as exclusive of all others. This, in turn, implies an abbreviated hierarchy of theoretical approaches, all of which are derived from a shared cosmologically defined knowledge system. On a meta-theoretical level and in terms of the corpus of contending theories, then, each individual approach is necessarily inclined toward synchronic, theory-displacing change as opposed to diachronic, cosmology-displacing change. That is to say, the diacritical mode of knowledge production orients critical approaches to usurp the seat of the hegemonic discourse rather than to unsettle hegemony itself. When they do so, they play a game of hegemony's making thereby reproducing the operant cosmology.

Its frame of reference restricted, the common usage of the term "counter-hegemonic" thus serves to obfuscate actual articulations of hegemonic power. I therefore propose a distinction between *contra*-hegemonic (sympathetic) and *counter*-hegemonic (empathetic) theory. The former is sympathetic in the sense that it advocates for emancipation, but on its own terms; the latter is guided by an empathetic ideal it can never truly realize but that nevertheless works to unsettle the hegemonologue in all its manifestations. If empathy means seeing the world through the eyes of the Other, *différance* and the prismic distortions of intersubjectivity are its eternal foils.[43] As an ideal, however, its own limits become its greatest virtue, leading us once again to the realization that our first commitment must be to conversation. An empathetic approach must readily admit of cosmological possibilities that are radically different from our own—this would be an approach that not only acknowledges the authority of marginalized voices but also necessitates that they be engaged in conversation. Only through the union of conversation with the empathetic ideal can the matrices of the intertext be traced; only thus can we confront our complicities with the hegemonologue and, through it, with advanced colonialism. Meanwhile, the concept of "contra-hegemony" connotes a challenge that is limited in its scope to fields of contestation bounded by the confines of the dominating knowledge system. And the neologism "contra-hegemonic" makes present—indeed, makes definitive—the violent absences enforced by the unproblematized use of "counter-hegemonic." For its part, "counter-hegemonic" is here assigned a much more expansive terrain and is revealed to be something we have yet to see in International Relations.

Insofar as international theory does not extend to raising challenges to the operant Western cosmology that both constrains and conditions it, it cannot rightly be regarded as counter-hegemonic, as I have defined it here. As contra-hegemonic theory and being inseparable from Western cosmological commitments, international theory implicitly accepts the conceptual terrain defined by hegemonic cosmological commitments and, therefore, contributes to the hegemony and exclusivity of Western cosmology by participating in its (re)production and propagation) The material consequences of this obtain in the circumstance that those who demur from the supposed universal truths derived from the operant cosmology are apt to be cast as heretics or fools. There is thus a powerful proselytizing impulse inherent in these commitments that obviates the validity of Other cosmologies. And this impulse may be found at the root of the culturally assimilative programmes and practices carried out by the agents of early and advanced colonialism—most prominently on behalf of the European empires, a variety of Western religious institutions, and the settler state—toward the extinguishment of the traditional worldviews and lifeways of Indigenous people(s) in the Americas and elsewhere. But impersonal vehicles, such as the fundamentals of Western science (especially as articulated through social theory) and the spread of Western material culture (bearing the knowledges of colonialism/advanced colonialism), have also functioned to advance the broad implantation of Western cosmological commitments.

A traditional Lakota worldview operationalizes an appreciation of and sensitivity to cosmological diversity that is a prerequisite to the development of an empathetic approach to theorizing the international. In the most fundamental sense, the point to which the above discussion brings us is that an idea like "security," for example, cannot properly be rendered as an unqualified good. As a concept, it must not be treated as monolithic such that it implicitly denies and defies a complete range of alterity. The point here is not only that different conceptions of what constitutes security as well as a variety of propositions regarding its appropriate referent object(s) cannot be accommodated by extant theoretical approaches. Rather, it is that this implicates International Relations in the violences of advanced colonialism, even in its more critically inspired articulations—these violences are aspects of its deferred meaning that we must work to make present. And the point is also that international theory is impoverished by a restrictive—if unseen—cosmology that does not admit of a full range of imaginable possibilities. Here, the corollary to the observation that international theory manifests as an advanced colonial practice is that precisely the same cosmological commitments that give rise to this situation exert a tyranny over the process of theorizing within the field itself to the extent that they both unduly limit our imaginings of the international and subvert the development of a broadly emancipatory project. These, together with some suggestions as to how we might begin to address them, are the issues that preoccupy me in the chapters that follow.

Notes

1. Exactly what that duty might be had never really been entirely clear to me, but the comic book seemed to aver big and important things.
2. See, e.g., Luce Irigaray's (1985a: 106–18) discussion of fluid mechanics and Paul Ernest (1998) on the social determinants of mathematical knowledge. More generally, see Harding (1998).
3. See also Ward Churchill's essay, "White Studies: The Intellectual Imperialism of U.S. Higher Education" (Churchill 1995).
4. See also, Foucault (1978: 92–98).
5. Even these, however, are easily complicated by the idea of a "middle."
6. Derrida holds that "in a classical philosophical opposition we are not dealing with the peaceful coexistence of a *vis-à-vis*, but rather with a violent hierarchy. One of the two terms governs the other (axiologically, logically, etc.), or has the upper hand" (Derrida 1981: 41).
7. In his *The Post Card* (1987), e.g., Derrida reverses the speaking positions of Plato and Socrates. And in "*La pharmacie de Platon*," he works the idea of the pharmakon, meaning both "cure" and "poison," into a new concept that is neither (Derrida 1972: 71–197).
8. See Derrida (1973: 129–60).
9. This, according to Derrida, is an inescapable aspect of presence. As he writes in *Positions*: "Whether in the order of spoken or written discourse, no element can function as a sign without referring to another element which itself is not simply present. This interweaving results in each "element"—phoneme or grapheme— being constituted on the basis of the trace within it of the other elements of the chain or system. This interweaving, this textile, is the *text* produced only in the transformation of another text. Nothing, neither among the elements nor within the system, is anywhere ever simply present or absent. There are only, everywhere, differences and traces of traces" (Derrida 1981: 26).
10. In his deconstruction of the speech/writing binary (an oppositional construction that privileges the former pole), Derrida works first to reverse the hierarchy by showing writing to be more proximate than speech to a present(able) truth. *Différance*, distinguishable from "difference" in writing but not as a spoken word, is thus an always and immediately tangible example of the indeterminacy of speech (Derrida 1973: 131).
11. The paradox of this retroactivity is that "[t]here was no signer, by right, before the text of the Declaration which itself remains the producer and guarantor of its own signature" (Derrida 1986: 10). Similarly, Campbell (1998) reveals how "history" has been retroactively made and deployed in the present so as to legitimize/explain aspects of the Bosnian civil war.
12. One need only speak imprecisely of "violence" to a group of undergraduates in order to appreciate the effect of this.
13. Isaac Newton could conduct his experiments with light refraction fully aware of the mediation by the prism: "In a very dark Chamber, at a round Hole, about one

third Part of an Inch broad, made in the Shut of a Window, I placed a Glass Prism, whereby the Beam of the Sun's Light, which came in at that Hole, might be refracted upwards toward the opposite Wall of the Chamber, and there form a colour'd Image of the Sun" (Newton 1952: 26). Newton knew both the prism and the light independently before he brought one to bear upon the other. In the interplay of subjectivities, however, the mediations of the "prism" may go unnoticed since we have no experience of any signified independent of external signifiers. The effect is as if two observers attended Newton's experiments, neither knowing whole light, but each thereafter comfortable speaking of "light" unaware that the other has seen it differently. Together they would share a signifier, but not a meaning of the signified.

14. Admitting to several deliberate misuses of concepts and theories from physics and mathematics in his original article, which he claimed would have been obvious to an undergraduate physics or math major, Sokal chided that "[e]vidently the editors of *Social Text* felt comfortable publishing an article on quantum physics without bothering to consult anyone knowledgeable in the subject" (Sokal 1996b: 63).

15. This is the order in which they are treated in dedicated chapters. Others, like Derrida and Jean-François Lyotard, are indicted along the way (Sokal and Bricmont 1997). An English-language version was subsequently published as *Fashionable Nonsense: Postmodern Intellectuals' Abuse of Science* (New York: Picador, 1998).

16. In a similar context, R.B.J. Walker notes of a common indictment of poststructuralism that it "is invariably issued in the name of objectivity and universal standards, although it is the historically constituted nature of the capacity to issue the indictment in the first place that post-structuralism has sought to challenge" (Walker 1993: 96–97). In any event, as Jim George points out, "The charge of normativism or subjectivism . . . has a pejorative sting only if one has already framed the pursuit of . . . knowledge in objectivist terms: in other words, if the normative or subjective is represented as the negative side of a subject/object dichotomy" (George 1995: 208).

17. The reference to "mysticism" here is inspired by their sense that Lacan's writings incorporate an aesthetic design that seems to imbue them with the qualities of sacred texts, stimulating reverent exegesis in the faithful: "ils servent de base à l'exégèse révérencieuse de ses disciples" (Sokal and Bricmont 1997: 39).

18. See Irigaray (1985b: 11–129).

19. "Relier le rationnel et l'objectif au masculin et le subjectif et l'émotif au féminin, c'est répéter les pires poncifs machistes" (Sokal and Bricmont 1997: 113).

20. Here Weightman is also responding to Irigaray's argument that fluid mechanics are underdeveloped because of their devaluation through association with the feminine.

21. Interestingly, this comment seems to undermine Sokal's case somewhat, betraying some of the very concerns raised by postmodernists and poststructuralists. After all, if rationality and objectivity do indeed prevail in the sciences, the political

inclinations of the scientists should scarcely be at issue. If, on the other hand, he means to suggest that, though it might be rational and objective, science will be selectively deployed (that is, brought to bear on some problems to the exclusion of others) according to the politics of the scientists, then he has accepted Irigaray's argument about the underdevelopment of fluid mechanics. Equally noteworthy, Sokal seems responsive to the matter of discursive authority and the relative (de)valuation of voice, not wanting to cede that privileged location he has constructed as the "intellectual high ground." Again, one is hard-pressed to see why it should matter who stands upon this "ground" so long as rationality and objectivity reign supreme.

22. Originating from very different sets of political commitments, see, e.g., Harvey (1990) and Gross and Levitt (1994).

23. See Linker (2001: 60).

24. Singling out Foucault's commitment to end injustice as an example (Hartsock 1987: 190).

25. Krishna is not charging deliberate complicity in any of this: "I am not suggesting that postmodernist analysts of the war are in agreement with the Pentagon's claims regarding a 'clean' war; I am suggesting that their preoccupation with representation, sign systems, and with the signifier over the signified, leaves one with little sense of the annihilation visited upon the people and land of Iraq" (Krishna 1993: 399).

26. See also (Campbell 1998: 173–75).

27. For Levinas, this view enabled a recovery of the politics inherent in the bystander's refuge in an ostensibly apolitical distancing of the Self from the violences visited upon the Other. Applying this idea to the war in Bosnia, Campbell finds that "there is no circumstance under which we could declare that it was not our concern" (Campbell 1998: 176).

28. To be clear, the problem is not with poststructuralism per se, but with the incompleteness of the hermeneutic circle. It is this circumstance upon which a more explicitly oppositional politics must be brought to bear.

29. By this I do not mean to evoke the linearity of a dialectical history. Rather, returning to the wave metaphor of a performative and inherently *current* history, I am interested in a theory of the ideational and material bases of our present moment; a theory of the wave itself and what sustains it in the present moment, or that only moment in which it has being.

30. That is, they seem to readily unsettle oppositional and status quo politics alike.

31. This, of course, is not to suggest a full and complete transition from structuralism to poststructuralism in the sense that the latter is the "after"-condition of the former. Rather, it is to say only that structuralism enables poststructuralism; structuralism is the precondition of poststructuralism, but they coexist in time.

32. Just as structuralism is preconditional to poststructuralism, modernism is the precondition of postmodernism. The idea of a "postmodern era" thus defines a substantial span of time (an "era") by a European referent. And, failing any specific

spatial qualifications, it is invoked for all places (and peoples) as though some undifferentiated "modernism" was once present everywhere. This at least marginalizes any account of an independent historical condition. What Kate Manzo describes as a "postmodern attitude" is considerably less problematic, more properly treating postmodernism as "a countermodernist attitude" (Manzo 1997: 386). While this is less suggestive of the ubiquity of modernism, it too gets tricky outside of European contexts inasmuch as it holds modernism and its effects as defining features of the operant "attitude." This leaves out of consideration the possibility of Indigenous practices (whether material or ideational) that, though they may run counter to modernism, are not born of it. The danger in this is that the Other might be inadvertently denied independent agency by a collapsing of the Other's practices into the denigrated pole of an action/reaction binary that privileges the modern European Self. Manzo is careful to disavow any such move, pointing out that "African or other 'third world' political thought with a more than passing resemblance to contemporary French philosophy is [not] merely derivative of it" (Manzo 1997: 403). But not all theorists are so conscientious in issuing this caveat, and where it is not deliberately advanced neither is it always self-evident.

33. For a thorough survey of postcolonial theory that traces its main currents from its origins up to and including current debates, see Loomba (1998).

34. See also Said (1993).

35. In an otherwise sympathetic critique of the Gramscian-inspired Subaltern Studies Group, Spivak argues that violent essentialisms—effecting the erasure of women's voices, for example—are an inescapable part of any attempt to recover a subaltern voice (Spivak 1988a).

36. See his essay, "Signs Taken for Wonders: Questions of Ambivalence and Authority Under a Tree Outside Delhi, May 1817," in Bhabha (1994).

37. For an extended psychoanalytic argument locating both an inferiority complex on the part of the colonized and a concomitant desire to be "white" amongst the tangible legacies of colonialism, see Fanon (1967).

38. The first articulation of Spivak's notion of "strategic essentialism" is found in an essay entitled "Subaltern Studies: Deconstructing Historiography," in which she describes it as a "*strategic* use of positivist essentialism in a scrupulously visible political interest" in order to raise a "collective consciousness" (Spivak 1988b: 205; emphasis in original).

39. For similarly inspired approaches see White (2000) and, drawing on White among others, Bleiker (2003).

40. Pratt defines the "contact zone" as "social spaces where cultures meet, clash, and grapple with each other, often in contexts of highly asymmetrical relations of power" (Pratt 1991: 34).

41. Ashcroft, Griffiths, and Tiffin offer that "[m]ore than three-quarters of the people living in the world today have had their lives shaped by the experience of colonialism" (Ashcroft et al. 1989: 1). I would argue, however, that the effects of colonialism

have extended even further than this. The hegemonologue has reached into locales never put under the direct authority of the European empires. Moreover, many of those same locales were affected by European encroachment when, for example, it interfered with established trading regimes or was otherwise disruptive of traditional lifeways. And, as we shall see in chapter 6, these effects were felt even in cases where Indigenous peoples and Europeans neither came into contact nor had any apparent awareness of one another's existence.

42. The critique of neorealism in International Relations, for example, has frequently involved charges of its complicity with U.S. global hegemony.

43. It is in this sense, and for reasons that become clearer in chapter 3, that I mean something different from what Christine Sylvester describes as "empathetic cooperation" (though I think in spirit the two ideas have much in common). This, according to Sylvester, is a stance that allows "analysts of IR" the opportunity "to think theory from strange empirical standpoints" (Sylvester 1994: 334). My usage of "empathy" is simply to describe an ideal orientation distinct from "sympathy." I would also want to distance myself from any implication that Others' knowledges simply inform our theorizing.

CHAPTER 2

Disciplinary International Relations and Its Disciplined Others

This book has, since its inception, been haunted by three conflicting propositions: that it possibly cannot, perhaps should not, and yet must be written. Each in its turn, these three frets speak to the issues to be taken up in this chapter and the next. As to the first, competency in the area of ethnographic research and writing methods was clearly requisite, but disciplinary International Relations has little to offer in this regard, making it impossible to proceed without a considerable investment of time and effort to become acquainted with the vast literatures on ethnography generated by (principally) anthropologists and sociologists. Second, and related to this, are the myriad ethical issues that unavoidably attach to any project involving ethnographic representation—particularly when undertaken from a position of privilege relative to those (re)presented. This has, quite rightly, given considerable pause to deliberate upon the moral quandaries that have been my constant companions throughout. But it does not necessarily follow from these daunting challenges and disquieting perils that the project cannot or should not be pursued. On the contrary, ethical dilemmas are not averted by respecting the disciplinary boundaries that have both isolated International Relations scholars from important debates about ethnographic research methods and simultaneously underwritten their inattention to Indigenous peoples. As we shall see in the chapters that follow, these omissions have, in part, implicated disciplinary International Relations in processes of advanced colonialism by helping to reproduce the hegemonic ideas, categories, and narratives that have enabled historic and ongoing domination over

Indigenous peoples. Thus, the choice not to proceed, while promising deliverance from the necessity of onerous extradisciplinary engagements, would also amount to an acceptance of a highly inequitable status quo of power relations and as such cannot be read as a neutral stance.

What this immediately points up is the imperative that postcolonial international relations be theorized without deference to the limiting conventions of disciplinary International Relations. Disciplinarity itself is inseparable from constitutive accounts of Selves and Others that inextricably bind the disciplines to the enduring legacies of colonialism. With its institutional roots in the heyday of European conquest and expansion, academic disciplinarity reflects cultural and racial ideologies that endorsed the division of human societies into a hierarchy of types, assigning to each its own ostensibly apposite mode of academic inquiry. The emergent disciplines, of course, developed their paradigms and methodologies in accordance with the assumptions of their founding. Thus, the invisibility of Indigenous peoples in International Relations, following from authoritative dictates as to the appropriate subject matters of the field, is also reproduced by innate practical impediments to unconventional (from a disciplinary standpoint) research agendas. While it has been clear from the outset of the present project, for example, that ethnographic research would be necessary, International Relations does not train its students to undertake this kind of inquiry in fieldwork settings. Nor does it equip us to distinguish good ethnographies from bad ones—a prerequisite for much of the textual analysis undertaken in later chapters. Methodologically ill prepared to do the kind of empirical work often called for when engaging nontraditional subject matters and lacking the competencies needed to discriminate between existing ethnographies, students of International Relations do not easily breach the boundaries of the discipline to expose its advanced colonial complicities.[1]

It is in this sense that the methodological issues that, *pro forma*, must be outlined in any book like this one, are here an integral part of the thesis itself. If International Relations is implicated in the maintenance and reproduction of advanced colonialism, this has much to do with the disciplinary parceling of knowledge realms by which certain categories of people are denied the possibility of an "international" presence. Students of International Relations lack competency in ethnographic research methods precisely because these methods have been developed with reference to peoples whose political lives are seldom imagined to be relevant beyond the domestic contexts of the states now mapped over their environs[2]—an enduring legacy of colonialism. Put another way, ethnography has largely been regarded as the knowledge-gathering technology appropriate to researching peoples whose voices are not otherwise spoken by states of their own. People or peoples, then, are the referent objects of ethnographic research, in contrast to the states that have been the traditional focus of International Relations. Of course, this is not to say that students of International Relations have not frequently consulted human repositories of the knowledges they have sought to uncover through their research activities. These informants, however, have typically been engaged for reasons of the offices they hold and as such have been treated more as conduits through which the researcher might access the true object of study: in most cases, the state. Accordingly, they have been enlisted as expert witnesses to matters of interest, though without being imbued by their interviewers with independent ontological significance for the study of international relations.[3]

But as a growing number of International Relations scholars have begun to move beyond the old referents, a new body of fieldwork that is more ethnographical in character has emerged. More and more, informants themselves are of central interest to researchers. Activists, Indigenous people, migrant workers, and a host of others, increasingly recognized as much more than mere repositories of specific experiential knowledges, are now regularly approached not only for reasons of what they know but also out of an interest in and appreciation for their often radically different ways of knowing. Although these developments have by no means been embraced by the whole of the field, critically inclined International Relations scholars, in particular, have sought by these investigations to unsettle many of the ontological and epistemological commitments of the orthodoxy of the discipline and, frequently, to advance some emancipatory project in the process. Their investigations have thus opened up possibilities for the epistemic enlargement of the field, making way, in turn, for ever more ethnographically based projects.

But while the epistemological terrain of the discipline has been steadily expanding in recent years, precious little attention has been devoted to its methodological dimension which, consequently, has not kept pace. Notwithstanding that two decades have passed since James Clifford (noting the cross-disciplinary proliferation of research projects dealing with issues of culture) was already able to comment that ethnography had become "an emergent interdisciplinary phenomenon" (Clifford 1986a: 3), associated theoretical work remains little known in International Relations, and even less so in Security Studies. To be sure, there have been a few very promising moves toward confronting these sorts of issues from within International Relations itself,[4] but to date no book-length attempt to think through the emergence of ethnographic research in the field has been published and few inroads have been made in terms of curricula, as even a cursory survey of course syllabi will attest. Such prefatory engagements as have been made are most frequently, as here, subsumed within a larger (usually critically oriented) project, so that the student of International Relations might be forgiven for thinking them to be rather idiosyncratic. The unsatisfactory result of this critical lapse is that researchers embarking on unconventional projects are left to guess at the elements of an appropriate methodology. Treading paths traditionally reserved in the main to anthropologists and sociologists, they may want for any acquaintance whatsoever with the lively and exhaustive debates about ethnographic research and writing that have been underway for decades in those disciplines. Even more troubling, the lack of emphasis on research methods in International Relations more generally might give some researchers to assume that such considerations are unimportant and thereby to miss seeing the myriad ways that problematic methodologies could frustrate the objectives of their research, or worse: as we shall see, interpretive problems, ethical considerations, and the danger of "colonizing" informants' knowledges are but a few such pitfalls.

Most fundamentally, it is by interrogating the origins of disciplinarity itself that we begin to reveal the varied collusions of disciplines in the advanced colonial domination over Indigenous peoples. A discussion of extradisciplinary debates about ethnographic research and writing is thus doubly at the service of the broader aims of this book. First, it calls for a critical examination of the disciplinary precepts by dint of which certain modes of inquiry have been deemed appropriate only to the study of particular kinds of peoples. And it obliges us to engage extradisciplinary literatures that explore the limits of ethnographic representation, enabling the revelation that even the most honestly conceived of the hegemonic accounts of our Others (to say nothing of our accounts of ourselves) are highly suspect. Given the historical relationship between disciplinarity, ethnography, and colonialism, the context in which the disciplines emerged in the Western academy speaks directly to the methodological considerations that have guided the present project. Indeed, though it is addressed primarily to an audience in International Relations, the whole of this project is self-consciously undisciplinary, such that this is properly counted among its methodological commitments. It is therefore appropriate that we turn our attention to the matter of disciplinarity before introducing some vital issues concerning the production of ethnographies in chapter 3, at the end of which some ethical problems are addressed in light of both the literature on ethnographic research methods and the ongoing practices of advanced colonialism. The reader is reminded, however, that none of these amounts to an independent sojourn or digression—on the contrary, all of what follows in this chapter and the next is offered in support of and continuity with the line of argument being developed throughout the whole of the book even as it fulfills the more mundane requirement that methodological commitments be clearly specified.

Confining Knowledges, Defining Worlds

For many, the aftermath of the abrupt and largely unanticipated evaporation of its Cold War context has marked something of a crisis for disciplinary International Relations, which has since been left casting about for a new footing absent the circumstances that attended its most intensive period of institutional growth and development. So it is that the years since this watershed have seen considerable ink spilled and seemingly no end of introspective inquiries undertaken in myriad efforts to come to grips with how the study of international relations has been or should be (or, depending upon one's perspective, whether it has been or should be) changed by (or in response to) the transition to a post–Cold War world. Importantly, however, these developments are not everywhere experienced in negative terms as a moment of crisis. On the contrary, they have been very much welcomed in some quarters where the transformative possibilities they seem to signal are most heartily embraced. The sudden unsettling of disciplinary consensus about the purpose and orientation of International Relations also effected an opening for voices of dissent already in evidence in the 1980s, most particularly in the writings of feminists and poststructuralists.[5] Besides raising a host of new challenges to the ontological foundations and epistemological commitments of the mainstream of the discipline, these new dissident[6] strands of scholarship began to destabilize disciplinary practices by denying them the certain ground(s) upon which their claims to an exclusive academic domain were founded. As Richard Ashley and R.B.J. Walker put it:

> [D]issident works of thought . . . accentuate and make more evident a sense of crisis, what one might call a crisis of the discipline of international studies. They put the discipline's institutional boundaries in question and put its familiar modes of subjectivity, objectivity, and conduct in doubt; they render its once seemingly self-evident notions of space, time, and progress uncertain; and they thereby make it possible to traverse institutional limitations, expose questions and difficulties, and explore political and theoretical possibilities hitherto forgotten or deferred. In short, dissident works of thought help to accentuate a disciplinary crisis whose single most pronounced symptom is that the very idea of "the discipline" enters thought as a question, a problem, a matter of uncertainty.   (Ashley and Walker 1990b: 375–76)

Writing from the margins of the discipline themselves, the dissidents' own struggles for legitimacy as theorists of international relations could scarcely help but strike, at least implicitly, at the foundations of disciplinary boundaries.[7]

What this means is that attempts to broaden the contemplative agenda of the field have not just been *met* with but have in fact been *one* with struggles around the practice of disciplinary "gatekeeping." A sometimes heated debate about the appropriate boundaries of Security Studies has, for example, resulted from attempts to treat the biosphere as a referent object of security.[8] More generally, the increasing number of dissident voices has occasioned a vigorous policing of the field by those who would have such interlopers apprehended and banished for the impertinence and impropriety of their willful disregard of disciplinary convention. While this reaction has been quite acutely felt throughout International Relations, it has been especially pronounced in the sub-field of Security Studies.[9] Here, as with International Relations as a whole, a number of the telltale traits of a discipline have long been exhibited, notwithstanding the fact that neither of these realms is usually endowed with formal disciplinary status.[10] In particular, there is near consensus about the appropriate objects of study and a more or less agreed body of core concepts articulated through a specialized vocabulary. This is backed up by an endogenously developed literary tradition that includes a number of *sui generis* "classics"—required reading on most course syllabi—and a reasonably large complement of specialized journals. The field is also characterized by a high degree of institutional isolation, reflected most vividly in the failure to develop conversations with scholars in other disciplines around a security studies specialty.[11] Finally, attempts to broaden the empirical and conceptual reach of Security Studies have been answered with unambiguously disciplinary calls for purity.[12]

While there are relatively few widely acknowledged disciplines in the social sciences—numbering perhaps a dozen[13]—the formal definition by which they come to be recognized as such has tended to render enclosed spaces, imagineering[14] disciplines as "sovereign territories" whose internal coherence is a function of shared subject matters, research methodologies, and paradigms. But taking these attributes as given, as though by nature, misconstrues disciplines as things instead of practices—to the extent that they are routinely treated as though they had corporeal presence, they have been constructed at the level of ontology. And this implies a certain stasis that, transhistoricizing claims about the objective content and limits of a given discipline, lends well to gatekeeping and policing practices. This is reflected in our apparent inability to talk about disciplines without recourse to spatial metaphors: they may be described as "islands" or "atolls" (Nissani 1997: 202, 211) or defined by the "borders" (Dogan 1997: 433) that separate them; those who move from one to the next are therefore "immigrants" (Nissani 1997: 201, 205) in "exotic lands" (Nissani 1997: 211) or brokers of the "trade routes" that sometimes traverse disciplinary "frontiers" (Dogan 1997: 433); these travelers will no doubt be grateful for the odd "road map" or reference to a "landmark" given by those who would wish them well in their visits to "neighboring" realms (Holsti 1989: 16); disciplines themselves are occasionally even imbued with agency such that each "jealously guards the sovereignty of its territory" (Dogan 1997: 429). They are thus imagined in ways that bespeak stable and readily apprehendable insides and outsides. Dissident discourses are then easily—if problematically—marginalized by way of reference to secure definitions of the terms and limits of legitimacy in the discipline.

Equally problematic, the ontological rendering of disciplines in spatial terms mystifies their temporal contingency. Disciplines are not timeless things; indeed, they are not "things" at all. Though we might often think of them as sets of limits, this is to confuse them with their effects. Disciplinarity is more properly understood as the sum of particular *practices* rooted in an epistemological stance that sanctions the parceling of ostensibly discrete knowledge realms. Disciplines can only be said to "exist," therefore, to the extent that the practices of disciplinarity persist. In this sense, to speak of a "discipline" is somewhat misleading—"discipline" is a noun, and nouns ontologize. We do not encounter disciplines in the way we would expect to come upon something with ontological presence. Rather, we *experience* them in the manner of any instance of our having been acted upon. "Discipline," then, needs be taken as a shorthand for an amalgam of the disciplining practices to which we and our work are subjected. This view moves us to see disciplines less as sets of limits and more as exercises of control. As Julie Thompson Klein describes it:

> The underlying action of disciplining knowledge is control. Control extends across the entire system of disciplinary technologies, from the structure of the curriculum, organization charts, and knowledge taxonomies to choice of dissertation topics, decisions about tenure and promotion, and judgments about publication and the awarding of grants. Disciplines control problems by naming the things that will be attended to and framing the context in which they are attended. (Klein 1996: 140)

Understood in this light, disciplines are forever being (re)made through practices with which they are coterminous. Where they are treated as though they exist in a corporeal sense they can also be read as (improperly) ontologized epistemological commitments. But they are always contingent upon the persistence of the practices that make them discernable as institutionalized sets of limits in the first place. Disciplinarity must be performed in order to be.

It is thus that we can rightly treat both International Relations and Security Studies as disciplines in their own right. Though they might not be counted among the recognized disciplines, they nevertheless embody the defining features of disciplines understood as practices. As noted above, breaches of consensus on the acceptable range of discourses and subject matters beget policing just as metaphorical boundaries are secured by gatekeeping practices. Replicating and extending these functions, core journals are notoriously resistant to nontraditional concepts and cases. Both International Relations and (especially) Security Studies are consequently among the most insular realms in the academy, their literatures correspondingly among the most incestuous. By these indicators, both arguably evince much more coherently disciplinary dispositions than many of the recognized disciplines.

Both the seriousness of efforts to breach the metaphorical disciplinary walls and the tenacity with which they have, in turn, been defended belie the origins of International Relations as a surprisingly interdisciplinary area of study, much better described as a broad specialty—or perhaps the convergence of a number of closely related specialties—than a discipline. A genealogy of the field will find its roots in, among others, Economics, History, International Law, Comparative Politics, and Sociology.[15] And yet, utterly concealing the heterogeneity of its pedigree, International Relations has evinced such a highly developed degree of disciplinary pretension that much of its core literatures have become almost completely self-referential—so much so that Mattei Dogan has seen fit to comment that the field "has become, in research if not in teaching, a quasi-independent domain" (Dogan 1997: 432). Here too, this relative isolation is as readily revealed in a survey of course syllabi (especially at undergraduate levels) where few, if any, extradisciplinary forays are ever found among assigned readings. It is most patently in these expressions of an unequivocal disciplinary inclination, amounting to the practical disavowal of its origins as a field of study reaching across a number of disciplines, that international relations has become International Relations.

This does not mean that the field has at all times been blissfully parochial. As a relatively new discipline with fairly diverse conceptual origins, it can hardly deny its indebtedness to a range of "classics" that it cannot rightly call its own. Thus, while Martin Wight has quite famously lamented that there is no international theory (Wight 1966), Stanley Hoffman has been more inclined toward the view that, while the discipline has essential works, there are many among them to which it cannot lay proprietary claim—even the discipline-defining works, so indispensable that Hoffman would recommend them to the exclusion of all else, speak to International Relations from across disciplinary boundaries (or, at least, from sites of some disciplinary ambiguity):

> [I]f I were asked to assign three books from the discipline to a recluse on a desert island, I would have to confess a double embarrassment: for I would select one that is more than two thousand years old—Thucydides' *Peloponnesian War*, and as for the two contemporary ones, Kenneth Waltz' *Man, the State and War* is a work in the tradition of political philosophy, and Aron's *Peace and War* is a work in the grand tradition of historical sociology, which dismisses many of the scientific pretenses of the postwar American scholars, and emanates from the genius of a French disciple of Montesquieu, Clausewitz, and Weber. (Hoffman 1987: 14)

Occasionally, it might seem as though International Relations' disciplinary pretensions are unsettled, if only fleetingly, by reflections of this sort. Most often, these moments are born of work on core traditions that finds itself, paradoxically, sited at the interstices between disciplines—Benedict Kingsbury and Adam Roberts, for example, find that the separation of International Relations and International Law appears quite problematic when one begins to work through a Grotian tradition in international theory (Kingsbury and Roberts 1990). Even so, disciplinary International Relations is not significantly destabilized either by its extradisciplinary foundations or by the occasional reference drawing our attention to them. Instead, such observations are more apt to be read—and, indeed, to have been written—as expressions of its relative newness than as charges of disciplinary non-viability.[16] It is not insignificant that, in spite of the minor dissidence of their momentary misgivings about the disciplinary division of knowledge, Kingsbury and Roberts articulate their interest in Grotius through the idea of a Grotian tradition in International Relations—working from an unmistakably disciplinary standpoint, they unavoidably participate in its (re)production.

Certainly, it is also possible to point to conspicuous cross-disciplinary appropriations of a more contemporary nature. Even Security Studies, notwithstanding determined efforts to maintain its disciplinary purity, has often embraced conceptual infusions from elsewhere in the academy. A good many of these borrowings have been favorably received and a few have even become quite influential—the extensive borrowings from Social Psychology that have shaped the perception/misperception literature are a case in point.[17] But it is important to distinguish between bona fide cross-disciplinary engagements and mere appropriation. Most extradisciplinary contributions to International Relations and Security Studies are instances of the latter. That is, ideas and insights are, from time to time, appropriated from without, but seldom is there any sustained conversation with their sites of origin by which we might know how well they have withstood scrutiny. Here too, the epistemological commitments of disciplinarity are betrayed: in this instance it is a positivist inclination that is exposed by the apparent sense that the "truths" mined from another discipline have a timeless validity that absolves us of any need to check back to see how they are faring. There is thus no sense in which we speak back to the disciplines whence came our appropriations. Rather, a wholly utilitarian motive is at work, as Ole R. Holsti unwittingly makes clear in an overture to Diplomatic Historians whom he would like to have better acquainted with International Relations:

> The study of international relations and foreign policy has always been a somewhat eclectic undertaking, with extensive borrowing from disciplines other than political science and history. At the most general level, the primary differences today tend to be between two broad approaches. Analysts of the first school focus on the structure of the international system, often borrowing on economics for models, analogies, insights, and metaphors, with an emphasis on *rational preferences and strategy* and how these tend to be shaped and constrained by the structure of the international system. Decision-making analysts, meanwhile, display a concern for domestic political processes and tend to borrow from social psychology in order to understand better the *limits and barriers* to information processing and rational choice.   (Holsti 1989: 40; emphasis in original)

These are appropriations in the truest sense, made in order that the analyst's "toolkit" is supplemented. The borrowings themselves, however, are not independent objects of curiosity and therefore languish in an imposed stasis that would render them every bit as real tomorrow as they are presumed to be today. Here is the disciplinary subversion of the interdisciplinarian's hopes. And the disciplinary pretensions of International Relations remain well intact, surviving Holsti's admissions of heterogeneity as easily as they survived Hoffman's—and in much the same way.

Lapsing momentarily into the spatial metaphors of disciplinarity, we might think of these extradisciplinary insights and ideas as having been granted "worker's visas," but with little hope of ever enjoying the rights of full disciplinary "citizenship." That is, they have been allowed a conditional admittance that stops short of legitimating them as genuine objects of inquiry unto themselves; within the confines of their adoptive discipline they are condemned to remain as found.[18] The practices of disciplinarity ensure their perpetual outsider status by continually reaffirming that every knowledge has an apposite disciplinary realm that is the exclusive domain of whatever specialized skills have been deemed appropriate to its advancement. But recognition of the temporal contingency of disciplining practices, by exposing the fundamentally performative essence of disciplines, highlights the matter of their historical complicities as well, so that this is exposed as a necessarily political move. After all, if disciplinarity resides only in the practices of human subjects occupying particular (typically privileged) moments/locations in sociopolitical time and space, it can hardly be unconnected to those contexts. The terms upon which extradisciplinary ideas and insights are, variously, embraced, conditionally admitted, or ruled out of order therefore cannot be read as politically neutral since disciplinary International Relations is a practice that inevitably results in the construction of privileged and marginal sites: in Foucault's words, "[d]iscipline is an art of rank" (Foucault 1977: 146).

That some discourses and subject matters are privileged while others are marginalized is, in important ways, reflective also of enduring ideas about the hierarchical division of humanity. Indeed, the very origins of the disciplines are inseparable from a confluence of social, cultural, and racial prejudices that, like the disciplines themselves, assign different spaces and imagine different limits upon the possible for various categories of people. Though commonly traced to the Enlightenment division of knowledge, the true origins of the contemporary disciplines are actually located in socio-evolutionary ideas dominant in the late-nineteenth century. The conventional disciplines we know today emerged in the period between 1850 and 1914 and, by the end of World War II, hardened to the point of virtual immovability (Wallerstein 1995: 840). Certainly, the division of knowledge, evident most readily in the radical separation of religious and logical-empirical spheres of knowledge production, was foundational to the Enlightenment. But such division as was undertaken in the early-modern period did not imply the raising of impermeable boundaries between discrete knowledge realms. That is to say, separation was not conceived as isolation. Rather, as Michael McKeon points out, these spheres were explicitly experimental and constituted with the expectation that their inherent interrelatedness would be upheld and that their value lay in the insights to be had at the interstices between them: "the famous Enlightenment projects in the division of knowledge often aimed, with greater or lesser explicitness, to provide thereby a new foundation for a unified scheme of knowledge" (McKeon 1994: 18). The emergence of disciplinarity and the ontological construction of discipline-things that began in earnest in the late-nineteenth century is thus better understood as a subversion of Enlightenment designs than as their fulfillment. Indeed, in McKeon's view this could only have taken place in a late-modern context: "the naturalization of modern categories was achieved only when their experimental and contingent constitution in the early modern period was sufficiently distanced as a historical phenomenon to be detachable from the categories 'themselves' " (McKeon 1994: 18).

But if disciplinarity is a post-Enlightenment phenomenon, in what historical particularities should we rightly seek the origins of the disciplines? Apropos of the spatial metaphors used to describe them, they are utterly inseparable from the apex of European expansion. In fact, the metaphorical territoriality of disciplinarity mirrors the territoriality of the colonial division of the world so neatly that it is fair to say that it is not entirely metaphorical. Disciplinarity first assumes the divisibility of the object. In theory, the object so dismembered in the disciplinary division of knowledge is human activity, parceled off into discrete spheres marked out as, among others, cultural, economic, or political. In practice, however, and betraying its colonial complicities, the object of disciplinary division has been humanity itself. Though ostensibly organized around realms of human activity, the founding of the disciplines was bound up with evolutionist notions that uneven social development between the various peoples of the world meant that not all participated meaningfully in the same realms. Where Europeans were endowed with economics and politics—both presumed to be marks of advanced civilizations—there were Others whose behaviors were reckoned to flow from cultural dictates alone. Hence, Economics and Political Science were founded with a geographic orientation quite distinct from that of Anthropology. Disciplinarity, it turns out, is an exercise in cartography.

It is important, however, to appreciate that the disciplinary division of knowledge—and, by extension, of humanity—is more than just a vulgar expression of racial ideologies. It is also an outcome of competing strategies toward the satisfaction of objectivist epistemological commitments. But neither these commitments nor the strategies devised for their fulfillment have ever been free of the forces and fetters of colonialism. On the contrary, they were necessarily conceived and advanced in accordance with dominant ideas about the evolutionary-hierarchical ordering of peoples. Significantly, there need be no malice or conscious machination involved here. In fact, there is usually little basis upon which to allege anything other than genuine efforts to advance the cause of knowledge. Instead, the pathological mappings of disciplinarity reflect uninterrogated commitments, even common senses, so deeply held that they may be taken as expressions of a discernible European worldview at a particular historical juncture. What we find here, then, is more in the way of a congenital entanglement of disciplinarity with the business of colonialism, not so instrumental a relationship as a straight functionalist account might suggest. And though it might sometimes be tempting to charge wickedness, such individualization of guilt mystifies the impersonal forces at work by way of the problematic implication that a few bad-hearted souls unscrupulously wrought tyranny from an otherwise benign order.

Preceding the dismemberment of the object was its isolation from the subject. This was a requisite move for objectivists, who sought to guarantee the purity of knowledge by excising the distortive interventions of an interested subject. Historians proposed to achieve this by way of a temporal separation, confining their investigations to events of the past. If subjectivity was to be banished, then the interestedness of researchers and informants alike was at issue. The passage of time seemed to address both concerns, removing the researcher from any immanent attachment to the matters under investigation (Wallerstein 1995: 842) while denying informants an active voice. In the case of the latter, subjectivity would be thwarted by a methodological bias toward archival material—archived documents were taken to be the most trustworthy sources of data, as they arguably reflected the most important events of bygone days and were presumed to be free of the embellishments and pretences feared from living informants who might sense the opportunity to shape a potentially influential account of events in which they have a stake (Wallerstein 1995: 841). Of course, this also meant that societies with oral literary traditions or, in any event, those that did not possess significant document archives did not lend well to study by historians. And here is the most profound sense in which, following Eric Wolf, Europe became distinguishable from those who were constructed, quite literally, as "the people without history" (Wolf 1982).

In the social sciences, a different strategy was employed. Choosing to work in the present, social scientists eschewed qualitative analysis in favor of quantification. This, it was thought, would make possible the application of scientific principles of investigation that the social scientists believed were better suited to the exclusion of subjectivity. Whereas historians looked to the past for patterns or trends that could be generalized across time, economists, political scientists, and sociologists sought to develop quantitative statements with universal validity. Economics, Political Science, and Sociology—respectively corresponding to the supposedly discrete realms of economics, politics, and civil society—thus adopted a nomothetic stance whereby they posited the existence of universal truths, worked to uncover and verify them in accordance with the rigors of science, and presumed to pronounce their timeless validity (Wallerstein 1995: 842–43). The social scientists reversed the methodological commitments of the historians, working from the present to make generalizations that could speak as well to the past. Like disciplinary History, however, the nomothetic social sciences seemed to their founders to befit the study of Europe more than its Others.

Where academic inquiry ventured beyond European contexts, specialized disciplines emerged. The influence of August Comte and Comtean positivism in the late-nineteenth century underwrote this development with the insistence that every science should have its own distinct subject matter. More particularly, though, Comte endorsed an evolutionary account of human social development, as reflected in his three stages of knowledge: theological, metaphysical, and positive. The first of these, according to Comte, corresponded to "primitive" societies, the second to "intermediary" ones, and the third to those that had achieved science (Neufeld 1995: 24). This coincided with the evolutionary-hierarchical idea of progressive stages of human social development. It also dovetailed with prevailing racial theories of the late-nineteenth century that, as Martin Bernal has argued, have been inveterately connected to disciplinarity since the very birth of the disciplines (Bernal 1987: 220). And together with these, it reserved the applicability of the three nomothetic social sciences to European contexts. After all, certain knowledge and universal truths, the very currency of these disciplines, could not be expected to follow from theologically or metaphysically based epistemologies. This put so-called primitive and intermediary societies beyond the pale of Economics, Political Science, and Sociology just as surely as did the imperialist certitude that economics, politics, and civil society were realms not fruitfully sought outside of European contexts.

Oriental Studies was established as home to those who would study societies judged to fit Comte's "intermediary" designation. Of interest here were the much-romanticized "high civilizations" of Asia that, one by one, had fallen under the control of the European empires. As Wallerstein has observed, disciplinary History was not easily brought to studies of "the Orient" as historians had little interest in what were assumed to be "unchanging despotisms" (Wallerstein 1995: 848). Similarly, the nomothetic social sciences seemed ill suited to scholarly engagements with supposedly static societies not endowed with the attributes of economics, politics, and civil society—at least not as those things were known in Europe. This lack of progressive economic, political, or social activity confirmed also that these were peoples without history. However, they were possessed of ancient texts and, in combination with the assumption of their relative stasis, this seemed to recommend a philological approach committed to the enduring relevance of writings from antiquity. The focus of Oriental Studies, then, was on texts. Denigrating though this might have been in its founding rationale, it did at least confer a conditional validity upon local voices as expressed through respectable literary traditions.

But if this was the approach appropriate to studying "intermediary" societies possessed of great texts, it would not do where no tradition of written literature existed. The so-called primitive peoples of the world were not imagined to have formed societies of the sorts known in Europe or Asia and were defined in the academy largely by way of negation: they were those who had no texts to be analyzed, who were determined in all respects by culture rather than by economics or politics, and among whom nothing akin to a bona fide civil society had emerged. And although historians sometimes devoted considerable attention to them—as, for example, where Indigenous North American people(s) were concerned—the assumption that these too were unchanging social worlds meant that they could only be treated as secondary objects of interest. That is to say, they became important to accounts of the dynamic Euro-American society with which their fates were inextricably intertwined, but were not normally imbued with independent significance. And as they were not imagined to have progressed to the evolutionary development of societies with differentiated economic, political, and civil society spheres, economists, political scientists, and sociologists showed little interest in them. Another specialized discipline seemed in order.

The need was filled by the new discipline of Anthropology and the methodological advent of formalized participant-observation. Centuries earlier, the first works of ethnographic description were produced not by scholars so much as by missionaries and colonial administrators who, having found themselves in regular and sustained contact with the "natives" they encountered as the European empires extended their reach throughout the world, presumed to speak of these strange (to Western eyes) peoples with authoritative voice. Seeing neither recognizable texts nor those attributes of advanced societies that would later become the foci of the nomothetic social sciences, these early ethnographers attempted definitive reportage by way of reference to culture alone. The authority of their accounts was conferred through pretensions of personal experience. In consequence, many of them averred much more than mere observer status vis-à-vis their informants, readily taking on airs of actual in-group membership. This conceit followed participant-observation into the academy and its new disciplinary home in Anthropology. All such pretense, however, was gainsaid by the concealed perspective inherent in the assumption that different analytical and descriptive strategies were here required because these engagements were with "primitive" peoples. There are thus important elements of continuity between the representational strategies employed in writings authored by the first literate Europeans in the Americas and what would later be held up as the professional anthropologist's principal research methodology—a correlation that is quite telling with respect to the ongoing advanced colonial complicities of disciplinarity.

International Relations as Ethnography

The objectivist strategies used in the study of the West effected a much-criticized artificial separation of politics, economics, and civil society, but this did at least confirm the existence of each of these dimensions. Elsewhere in the world none of these were seen to exist because the commitment to certain "universal truths" made it impossible to see any but an ontologically abbreviated range of accepted indicators of these realms. Presumed universal truths about human nature, for example, rendered the state as the sole legitimate expression of the political—absent the state, no advanced politics could be imagined to exist.[19] In this way, distinctly European—and profoundly Eurocentric—truth claims became the absolute arbiters of the degree of (mis)fit between any given society and different disciplinary discourses; in application, these same truth claims pronounced upon the "primitiveness" of various peoples, reducing their worldviews and lifeways to expressions of culture alone. And if culture was the overriding determinant of "primitive" societies, the tools of disciplinary History and Oriental Studies seemed no better suited to studying them than those of the nomothetic social sciences—objectivist epistemological commitments would have to be satisfied by some alternate means. Participant-observation, premised upon the separation of subject and object, thus became the accepted method of a new positive social science of culture. As an objectivist strategy, it was not only founded upon but it also reified the era-sures of disciplinarity by remaining true to them: even the anti-evolutionist scientific Anthropology inspired by Franz Boas in the early-twentieth century, though ostensibly concerned with the whole of humanity, remained confined (in practice, if not completely in theory) to the study of so-called tribal peoples. This, of course, was hardly surprising given that the commitments of positive science precluded the immersive study of one's own culture, and most anthropologists were Europeans or Euro-Americans. Accordingly, the constructed boundaries of the disciplines held fast and participant-observation was reserved to the study of "primitive" peoples, reconfirming their want of history, economics, politics, and civil society by the very fact of its agreed unsuitability to the study of these realms.[20]

Keeping in mind that disciplinarity is properly understood as practice, the persistence of both the disciplinary confinement of certain peoples to particular academic realms and the epistemological commitments with which these divisions were co-constituted is inseparable from broader processes of advanced colonialism. The disciplinary Anthropology associated with Boas followed the nineteenth-century evolutionist ethnographers in taking the ravages of colonialism as an imperative *raison d'être*: the traditional lifeways of many of the world's "tribal" peoples (and, in some cases, whole peoples themselves) were on the verge of disappearing and it became a most sacred mission of the anthropologist to secure accounts of them before they vanished forever. The effect of this was to mystify these researchers' participation (and, no less, that of scholars working in other disciplines) in institutionalized discursive structures that implicitly marked out certain peoples as "primitives" or "savages," constructing them, however lamentably, in the unalterable path of a relentless evolutionary or historical progress.[21] That is to say, Anthropological scholarship was itself directly implicated in the "disappearances" counted among its foundational rationales for proceeding. George Stocking, Jr. puts it thus:

> [A]lthough in etymology and in underlying problem orientation "anthropology" was about all of humankind, it tended in practice to be limited primarily to peoples who, stigmatized as "primitive" or "savages," were regarded as racially, mentally, and culturally inferior. From this perspective, then, Boas' anthropology "pure and simple" was less an embracive "science of man" than the residuary disciplinary legatee of the dark-skinned savage (or, in Boas' more generous terms, "less civilized") peoples of the world. Methodological and conceptual leftovers from the emerging human scientific disciplines, politically dominated, and culturally despised, they were commonly thought in fact to be "vanishing." In these terms, "anthropology" was not only historically constituted, but might even be historically delimited—and, therefore, in the minds of its proponents, all the more urgent.   (Stocking 1995: 941)

Notwithstanding that Anthropology has, since Boas' day, come to cast a much broader empirical net,[22] the assumptions of the discipline's founding live on in the continued exclusion of Indigenous peoples from other sites of knowledge production—not least, International Relations.

The disciplinary division of knowledge and knowledge production, a late-modern hallmark of the Western academy, is thus inseparable from ongoing processes of advanced colonialism in the Americas and elsewhere. International Relations and Security Studies are, by extension, identifiable as advanced colonial practices, rearticulating anew the sites of margin and privilege whence answers to the emergent big questions about things like security are proposed. And like all disciplines, they are productive as much as delimiting, reasserting the primacy of particular knowing subjects and their authority to construct their Others in opposition to idealized constructions of themselves. Importantly, and as Ashley and Walker have observed, it is this privileged subjectivity—founded in a modernist conception of reason as the key to security in mastery of an uncertain world—that is found at the root of both disciplinary violences and those of the state:

> [T]hrough reason, man may subdue history, quiet all uncertainty, clarify ambiguity, and achieve total knowledge, total autonomy, and total power. This is the promise implicit in every claim of modern "knowledge"—a claim always uttered as if by "man" and in the name of "man." This, too, is the promise that the disciplines of modern social science make—a promise of knowledge and power on behalf of a universal sovereign figure of "man" whose voice a discipline would speak. And this, as it happens, is the same violence that legitimates the violence of the modern state— the promise, inscribed in a compact with "man," to secure and defend the "domesticated" time and space of reasoning "man" in opposition to the recalcitrant and dangerous forces of history that resist the sway of "man's" reason.   (Ashley and Walker 1990a: 262)

In a sense, then, disciplinarity can be read as a defensive reaction to uncertainty. In practice, however, it manifests as more of a resistance to alterity in its correspondence with a "universal sovereign figure of 'man.'" This is the hegemonic subjectivity constitutive of and constituted by the disciplinary division of knowledge. To the extent that disciplinarity is defensive, it works to secure not only the metaphorical territories of knowledge production but also the privileged place and identity of its hegemonic subject(s).[23]

What this exposes is that, despite their mutual inattention to issues of ethnographic research and writing, both International Relations and Security Studies are actually inherently ethnographic enterprises: like all disciplinary practices, they participate in the ethnographic (re)production of modern Western "man." And, by their omissions, they speak implicitly of the nature and place of "primitive" peoples as well, sketching accounts of them in the very bases of their exclusion. It is therefore not the case that ethnography has ever truly been missing from these disciplines. But what have been absent are ethnographic reflexivity and a fuller range of ethnographic voices. Disciplinary International Relations and Security Studies have been decidedly monological, speaking the voice of a masculinized European subject but not admitting of other subjectivities (Walker 1997: 73). Again, this orientation cannot be rendered as value-neutral. On the contrary, it is an expression of the hegemonologue of advanced colonialism, inextricably bound up with the same hierarchical separation of the world's peoples as attended the disciplinary parceling off of knowledge realms.

An interdisciplinary response, proceeding from the largely ossified structures of knowledge division themselves, promises no deliverance from this. Instead, serious engagement with marginalized voices is called for as a means to interrupt and unsettle the hegemonologue in favor of more polyphonic specialties. And among those with whom we ought to begin to converse are Indigenous people. But if we are to avoid the pathological effects of appropriation discussed earlier, this must come in tandem with a reflexive sense of the limits of ethnographic representations of both Selves and Others. For, as we shall see, ethnographic accounts of Indigenous peoples do not always bear their voices.

Notes

1. Though conceived as multidisciplinary even in the earliest planning stages, the research for this book was drawn out over a much longer period than originally anticipated in deference to the need to become acquainted with essential literatures on ethnographic research and writing. But professional pressures to publish do not always admit of such extradisciplinary investments if they mean that the tangible results of research are thereby made to come at longer intervals. This is a very real and formidable structural barrier to the crossing of disciplinary boundaries, however artificial those might actually be.

2. I refer here to Indigenous peoples specifically.

3. This is not to suggest that it is at all unproblematic that traditionally oriented scholars regularly conduct interviews and otherwise undertake research projects involving human participants without the benefit of at least a minimal familiarity with the literatures dealing with issues related to ethnographic research projects in general, fieldwork methodologies in particular, and the ethical dilemmas inherent in researching and representing our informants and their knowledges.

4. For a rare dedicated treatment, see Whitworth (2001).

5. See, e.g., Ashley (1987); Cohn (1987); Enloe (1989); Klein (1988); Tickner (1988); and the essays included in Der Derian and Shapiro (1989).

6. See Ashley and Walker (1990a).

7. See Whitworth (1994: ix–xii, 1–7).

8. See, e.g., Deudney (1990); Dalby (1992); Levy (1995); Mathews (1994).

9. For a discussion and critique of some of the more common of these policing practices and the sleights-of-hand by which they are often effected, see Krause (1998).

10. The one qualified exception being those places where chairs have been established in International Relations or Security Studies.

11. A survey of extradisciplinary abstracts reveals work on security issues (variously defined) and the concept of security itself that is not familiar to students of Security Studies.

12. See, e.g., Walt (1991).

13. According to Mattei Dogan (1997: 430), who treats disciplines as clusters of specialties, "In the archipelago of social sciences, there are today, according to the

definition we adopt, between ten and fifteen formal disciplines, but hundreds of specialties, sectors, fields, subfields, interstices and niches."

14. This term, borrowed from Walt Disney, signifies the transformation of a work of the imagination into something that is, deceptively, experienced as "real."

15. For more on the origins and development of International Relations, see Olson and Onuf (1985).

16. Neither Wight nor Hoffman argue that International Relations is not viable as a discipline and Kingsbury and Roberts do not make this their project either.

17. In particular, I am thinking here of such works as Jervis (1968) and Janis (1982).

18. I hasten to add that this should not be read as an argument denying the possibility of escaping disciplinary contexts, although I am somewhat skeptical of inter-disciplinary projects that involve "imports" from other disciplines since these are all too often one-way transactions that do not give rise to sustained conversations. Noting the frequent failure to sustain such conversations, however, is not the same thing as saying that they are not possible; as Arabella Lyon has rightly observed in arguing against such a defeatist position, "disciplines are not nation-states with inviolate borders" (Lyon 1992: 592).

19. This point is developed extensively in chapter 6.

20. The discursive authority vested in the academy should not be underestimated here. When the academic consensus is—or appears to be—that economics or politics are not fruitfully studied in a given context, this carries considerable weight as a truth claim unto itself.

21. Boas himself did not subscribe to the evolutionary hypothesis of societal development, insisting instead that cultures were highly complex self-contained systems that had to be understood and appreciated on their own terms. This, in fact, placed him in opposition to the social engineering of assimilationists to the extent that no progressive stages of development existed to be urged on. But despite his cultural relativism, Boas did participate in the hierarchical ordering of societies, substituting the idea of "less civilized" peoples for the evolutionists' "primitive" designation. Cultural difference was attributed to historical circumstances. Still, Boas shared the belief that "less civilized" societies were vanishing. The equally troubling implication of this is that whole peoples should be expected to "vanish" not through their own eventual transformation to a higher stage of social development, but because of their inevitable displacement by the expanding societies of Europe. In place of an evolutionary one, the irresistibility of an historical "progress" thus foretold their demise with an equal measure of certainty.

22. See, e.g., the essays included in Marcus (1999). This is not to suggest, however, that disciplinary purists do not still exert a limiting influence through gatekeeping and policing practices (Clifford 1997: 58–64).

23. According to Eviatar Zerubavel, who regards disciplinarity as a symptom of the "rigid mind," "[t]he fact that in individuals as well as groups rigid-mindedness flourishes particularly during periods of acute identity crisis suggests that it is actually grounded in an overtly defensive stance towards the world" (Zerubavel 1995: 1096).

CHAPTER 3

Ethnography, Ethics, and Advanced Colonialism

The prejudices and pretensions of disciplinarity and, no less, their colonial complicities, are as discernible in ethnographic representation as in the disciplinary division of knowledge. Ethnographers are often given to referring to those they study as their "subjects." The unreflexive pretension to represent, however, confirms that they are nothing of the sort. The ethnographic voice of the participant-observer is an unsolicited surrogate voice by the knowing (Western/scholarly) subject on behalf of unspeaking objects. The very rationale for participant-observation—as opposed to philology, for instance—presumes the voicelessness of those under study. And this puts the lie to their identification as subjects of the ethnographies so produced—stripped of their subjectivity in deference to that of the ethnographer, they are thoroughly objectified from the outset. The possibility of a truly *counter-*hegemonic project is thus obviated by the inclination to represent the Other rather than to work toward the audibility of the Other's voice(s). The hege-monologue here frustrates the empathetic ideal, collapsing the multiple subjectivities of those being represented into the monolithic category of the Other, an objectified unsubjectivity that, following Spivak, cannot speak (Spivak 1988a).

The enduring colonial legacies of disciplinarity are implicated here. The effect is akin to the notional silence of the proverbial tree falling in the forest with nobody there to hear: the seemingly voiceless *do* speak but, having been denied the possibility of doing so within the academy, we can hardly expect to hear them from in here. Instead, we "hear" the voice of the ethnographer speaking of—and in place of—subjects-cum-objects. The project of ethnographic representation presumes that ethnographers must create texts about their subjects as the sole means by which others, short of undertaking their own field-work, might access knowledges by and about them. This discounts the value of texts preserved as oral literatures, containing and concealing the knowledges they bear except as converted by the ethnographer into written form. This is largely because oral literatures, seemingly more fluid and susceptible to manipulation by their bearers, do not sit well with objectivist epistemological commitments. But as Neta Crawford points out, we might do well to be a little less sanguine about the relative fidelity of the written form:

> Written texts are handy because they are semipermanent. But written "primary" texts are no more omniscient than oral histories; in fact, they may be less so. Written texts usually are inscribed by individual authors who rarely give us a sense of how widely shared their interpretations are. Even if widely shared, the written history is necessarily incomplete and reflective of a particular set of concerns and biases. In contrast, given the process of preserving and transmitting oral history, we know that more than one author was involved in shaping the account, for the generation of oral history is a public event, subject to public scrutiny and correction. (Crawford 1994: 351)

Of course, memory is socially constructed and therefore inseparable from the power dynamics of sociopolitical conditions and relationships.[1] But surely the same must be said of written texts with respect to the contexts both of their inscription and those in which they are read. In any event, as Howard Harrod reminds us, the texts of the dominating society are both oral and written, and the latter are no more immutable than the former:

> In large state-based societies, meanings are carried not only in oral traditions associated with founding predecessors but also in documents, such as constitutions, and in laws that are ultimately backed by force. Despite the fact that they are embodied in writing, these meanings are not immune to reinterpretation. Social change arises out of periods characterized by conflicts of interpretation that may issue in a new social vision, as well as new or reformed social institutions and practices. (Harrod 1995: 101)

It seems, then, that the objection that oral literatures are suspect for being impermanent and susceptible to being altered to reflect the subjective inclinations of their human repositories is not so sound as might hitherto have been imagined—we may as rightly reject written texts on the same grounds that are used to deny the validity of oral ones. As Angela Cavender Wilson points out, then, it is worth deliberating upon the odd circumstance that researchers who would not be forgiven for overlooking other sources that might be relevant to their work often seem able to ignore oral literatures with impunity (Wilson 1996: 3).

There is a conspicuous irony in the exclusion of oral literatures and the insistence upon transmogrifying whatever knowledges they bear into written ethnographic accounts. Specifically, this approach to author-izing knowledge ignores the well-established legitimacy of orality in the dominating society itself. As Frank Alvarez-Pereyre points out, the oral transmission of knowledge is commonplace in everything from musical instruction to apprenticeships in the trades: "The transmission of such knowledge does rely in part on technical manuals, but it also relies on oral transmission" (Alvarez-Pereyre 1992: 107). How ironic, at any rate, that the devaluation of oral literatures should be possible in the academy, the institution of Western society wherein the privileged place of the oral transmission of knowledge (from the lectern and in the conference room) is most deeply entrenched. To be sure, even here the written word continues to enjoy a more privileged place, but not to such a degree that it is imagined that students might come to knowledge without the benefit of direct tutelage. The institutional arrangements of the academy and of the formalized relations between professor and student attest to the validity, indeed to the centrality, of the oral production and transmission of knowledge in the dominating society.

Is it the absence of the medieval and hierarchically based trappings of the university, then, that has endorsed the exclusion/dismissal of non-Western oral traditions? Perhaps not directly, but the absence of these things does signal the sources of the present quandary: the exclusion of the oral literatures of Indigenous peoples arises from the same racialized conception of the Primitive as does the emergence of ontologized disciplinarity, and this manifests also in related convictions about the relative legitimacy of different ways of knowing. In the university, the legitimacy of oral transmission of knowledge is reserved almost exclusively to a small cadre of institutionally sanctioned authorities: the professoriat. Indeed, the lectern is among the most recognizable of the material expressions of authority over knowledge, a license reconfirmed by the spatial arrangements of the lecture hall. That the human repositories of Indigenous knowledges are, despite their having attained general assent as legitimate bearers of those knowledges in their own social contexts, not regarded as having similarly authoritative voices within the academy reflects an epistemological prejudice; it highlights the rejection of the possibility of an Indigenous authority that, in turn, is based on the prior refusal to concede the validity of Indigenous ways of knowing. In short, it is a rejection of these knowledges on the basis of the presumption that they are not sufficiently regulated, that they are undisciplined.

Authority is privately held in the Euro-American context, vested in particular persons. Credentials, both formalized and customary, confirm the status of the "expert" whose voice is privileged accordingly. Formalized credentials are institutionally conferred through the granting of degrees, awarding of tenure, and by having one's work published; the last of these reaffirms publicly the approval of one's peers, so that the cadre of academic "experts" participates directly in the (re)production of the authenticity of its own constituent voices. Less formally, authority is claimed and signaled through the everyday practices of disciplinary conventions. In cases where participant-observation has been undertaken in fieldwork settings, for example, it is customary for ethnographers to offer up an explicitly autobiographical component among the written reportage of their work. Whether included as part of the principal written product or published separately, personal narratives are an indispensable companion to the ethnographic text inasmuch as they tell the story of the fieldwork itself: the hardships faced and sacrifices made in order to live among "one's people," humorous anecdotes about such things as cross-cultural social gaffes, and, sometimes, accounts of how research designs came to be frustrated by the unforeseen. And as Pratt has argued, these companion texts are indispensable as a means to salvage the legitimacy of the ethnographic text from the paradox of participant-observation: a research methodology that explicitly harnesses the researcher's subjective experiences as (ostensibly) an honorary member of the society under study in order to satisfy objectivist epistemological commitments that disdain subjectivity (Pratt 1986: 33).

The personal narrative of fieldwork is simultaneously a declaration of authority to speak. Describing the researcher's labors in deference to the only accepted avenues by which "authentic" knowledge can be produced in particular disciplinary contexts, this component of the project lends credibility to the claims and accounts borne in the ethnographic text. The conventions of participant-observation thus confer the mantle of authority upon the scholarly voices that supplant those of the people under study—small wonder, then, that Indigenous communities are so often resistant to having anthropologists in their midst.[2] A telling aspect of personal fieldwork narratives and their advanced colonial complicities is that they so neatly replicate the terms of authentication that lent credibility to the travelogues of the first literate Europeans to join the frontiers of colonial conquest in the Americas and elsewhere. Moreover, the pretensions of the ethnographer—and of ethnography more generally—bring about the denial of collectively held and validated authority of voice, in part because the academy cannot easily apply its hierarchically constructed matrices of credentials, the means by which the authenticity of knowledge claims is judged. The role of the ethnographer, then, is to collect the knowledges held by local informants and transcribe them into a form recognizable in the academy. In combination with recent poststructuralist and postcolonial challenges that have begun to unsettle the established idea of ethnography as a process of translation,[3] this suggests that the ethnographer might be better understood as an ethno-scribe involved in an imperfect process of media conversion whereby communally held oral literatures are transformed into written texts—texts that are wont to lack in fidelity to those they are presumed to have replicated.

All the more problematically, this involves the appropriation of what would be understood as someone else's intellectual property were it otherwise audible within the academy. This is so inasmuch as the authenticity of the scholarly texts produced in the process is secured by way of authorial attribution, knowledge being validated through direct association with the credentialed "expert."[4] Put another way, the conventions of publishing and of copyright signal ownership by the ethnographer and/or the publishers of ethnographic texts, thereby divesting the subject-cum-object of any claim to author-ity. Indigenous knowledges are thus wholly appropriated and, much like the extradisciplinary appropriations of International Relations and Security Studies, become ossified accounts and claims divorced from their originating contexts. Moreover, the not insignificant discursive authority of the scholar resists attempts by Indigenous peoples to reclaim their knowledges and traditions or even to intervene against the violences of misinterpretation or decontextualization—the voice of the ethnographer not only displaces those upon which it is predicated, but also avers absolute license as the ultimate arbiter of Indigenous knowledges and traditions.[5] As Vine Deloria, Jr. describes the problem:

> The realities of Indian belief and existence have become so misunderstood and distorted at this point that when a real Indian stands up and speaks the truth at any given moment, he or she is not only unlikely to be believed, but will probably be contradicted and "corrected" by the citation of some non-Indian and totally inaccurate "expert." More, young Indians in universities are now being trained to see themselves and their cultures in terms prescribed by such experts rather than in the traditional terms of the tribal elders . . . In this way, the experts are perfecting a system of self-validation in which all semblance of honesty and accuracy is lost. This is not only a travesty of scholarship, but it is absolutely devastating to Indian societies. (Quoted in Rose 1992: 404)

And this endogenously conferred authority to represent is claimed and asserted in spite of the fact that participant-observation can never adequately access the cognitive context of the knowledges, traditions, and lifeways (re)presented in ethnographic writing—a limitation that is obscured by the declaration of authoritative/authentic voice asserted through accounts of fieldwork experiences.

The ethnographic project, understood as the production of "authentic" texts from and in place of "inauthentic" ones, is ethically problematic from the outset. Enabled by the circumstance that oral texts are not directly renderable as legitimate in themselves, the ethnographer seeks to capture Indigenous knowledges and (re)present them in written form, ostensibly liberating them from the imposed obscurity of orality. But the inaudibility of the voices whence these knowledges originated should be cause for some considerable concern given that the imposition of the subjectivity of the ethnographer is, as we have seen, rather less than credibly foresworn. And it is in recognizing this failure to reconcile the paradox of participant-observation—that is, the attempt to wed objectivist epistemological commitments in representation together with an explicitly subjective claim to authority—that the most fundamental dilemma of ethnographic representation is brought into stark relief: the intervention of the subjectivity of the ethnographer means that all ethnographies are unavoidably exercises in interpretation. Being neither translator nor mere scribe, then, the ethnographer is engaged not just in a substitution of voice, but also in a substitution of the text itself. This is, in simplest terms, a process of *re*-inscription.

Once again, it is important to emphasize that none of this need be the product of instrumental design. That is, we should not make the mistake of supposing that it must be some meanness of purpose or character that is at the root of the ethnographic distortion of Indigenous knowledges; nor can we easily attribute it to sloppiness or incompetence. Rather, it is simply that all ethnography is contingent and indeterminate—as Clifford Geertz has famously argued, it is necessarily an interpretive undertaking (1973: 15 and passim). This means that despite the best intentions of the ethnographer, the ethnographic text will inevitably be an imperfect re-presentation, this owing to *différance* and the limits of intersubjective understanding—as Alvarez-Pereyre puts it, "to grasp is partially to mutilate, and . . . all representation is transfiguration . . ." (Alvarez-Pereyre 1992: 113). Participant-observation might facilitate sharing in material practices, but this is not the same as sharing in cognitive ones (the cognitive must also be understood as practice, lest we open the door to essentialism). This means that one can never truly share even in material practices where the cognitive context that gives meaning to them is absent. Regardless, none of these practices can properly become lived "reality" for the ethnographer who, after all, gets to go home when the research is done—a luxury of choice and mobility that is often in stark contrast with the exigencies of life for one's informants (Stacey 1988: 23; Whitworth 2001: 157). Notwithstanding the pretensions of a subjectivity shared with one's informants, then, the ethnographic text is necessarily contingent upon constructed truth claims made by an ostensibly knowing subject who, paradoxically, lacks the capacity to know. Moreover, as Clifford observes, "all constructed truths are made possible by powerful 'lies' of exclusion and rhetoric," so that "[e]ven the best ethnographic texts—serious, true fictions—are systems, or economies, of truth" (Clifford 1986a: 7). And here too, it is important to emphasize that the "lies" to which Clifford refers do not amount to a suggestion that these accounts are deliberately perjurious. Rather, "[p]ower and history work through them, in ways their authors cannot fully control" (Clifford 1986a: 7).

For Clifford, ethnography is "a performance emplotted by powerful stories" (Clifford 1986b: 98). The ethnographic text can thus be understood as an allegoric tale: "Allegory prompts us to say of any cultural description not 'this represents, or symbolizes, that' but rather, 'this is a (morally charged) *story* about that' " (Clifford 1986b: 100; emphasis in original). What this exposes is the role of subjective referents—a universalized conception of human nature, for example—as arbiters of meaning as we struggle to come to an understanding of the unfamiliar. This, in turn, profoundly destabilizes the uneasy compromise between objectivist epistemological commitments and the surrogate subjectivity of participant-observation:

> What one *sees* in a coherent ethnographic account, the imaged construct of the other, is connected in a continuous double structure with what one *understands* [linear conceptions of being, for example]. . . . Strange behavior is portrayed as meaningful within a common network of symbols—a common ground of understandable activity valid for both observer and observed, and by implication for all human groups. Thus, ethnography's narrative of specific differences presupposes, and always refers to, an abstract plane of similarity.   (Clifford 1986b: 101; emphasis in original)

Significant, then, is how we construct similitude. If the ideational referents of sameness are constructs that neatly befit particular Euro-derived philosophical commitments but do not map easily onto other peoples' lifeways or sit well with elements of their cosmologies, then our representations of them will surely be distorted. Moreover, the tendency of ethnographic voices to displace those of their subjects-cum-objects exposes the inherent violence of this.[6]

Underscoring the problem, Clifford recounts what could well be the quintessential anecdote on the ability of ethnography to take sustenance from itself. During the course of an interview with a Mpongwé chief in Gabon, an ethno-historian inquired as to the continued significance of a number of religious customs and concepts recorded by another ethnographer, the Abbé Raponda-Walker, in the early-twentieth century. When asked about a word that was apparently unfamiliar to him, the chief left the room momentarily, returning with a copy of Raponda-Walker's compendium of local custom, which remained open on his lap for the rest of the interview (Clifford 1986b: 116). Amusing though this cautionary tale may be, the unequal power relations that attend fieldwork—and, ultimately, the writing of ethnographic texts—inject a sinister note as well. Here, the privileging of the voice of the ethnographer is unmistakable, and this highlights its colonial complicities to the extent that it can never be completely divorced from the worldview and deeper philosophical commitments of its origins—commitments that underwrote the colonial conquest and subjugation of the rest of the world. This is all the more problematic when the ethnographer's account becomes accepted as the authentic one.

Perhaps better than any other measure, the perfection of an advanced colonial system of domination is signaled by its capacity to (re)make its Others in ways consistent with its own logics rather than theirs. The result is that Euro-derived concepts, categories, and commitments are naturalized, while those of the Others remade in the colonial encounter are rendered implausible. It is thus that, as Cecil King has provocatively put it, Indigenous people(s) have "become imprisoned in the anthropologists' words" (King 1997: 116). That is to say, the authority of ethnographic voice—which, as in Clifford's anecdote, often has the power to displace the voices of Indigenous people even in their own communities—brings in tandem the straightjacket of Euro-American concepts and commitments, into which Indigenous worldviews and lifeways must be made to fit. As King points out, this manifests as much in the inability to express Indigenous concepts in the languages of the colonizer as in the Euro-derived predisposition to divide and compartmentalize: "The language that anthropologists use to explain us traps us in linguistic cages because we must explain our ways through alien hypothetical constructs and theoretical frameworks. . . . We must segment, fragment, fracture, and pigeonhole that which we hold sacred" (King 1997: 116). And the most immediate effect of this is that Indigenous people are often hard-pressed to recognize themselves in the authoritative accounts about them:

> I am Odawa. I speak Odawa, but anthropologists have preferred to say I speak Ojibwe. My language is an Algonquin language, I am told, and it is structured by describing things as animate or inanimate, so I am told. English definitions of the terms "animate" and "inanimate" lead people to think of things being alive or not alive. Is this how our language is structured? I think not. In Odawa all so-called inanimate things could not be said to be dead. Does animate then mean having or possessing a soul? Is this a sufficient explanation? I think not. Is the animate–inanimate dichotomy helpful in describing the structure of my language? I think that it is limiting, if not wrong outright. For in Odawa anything at some time can be animate. The state of inanimateness is not the denial or negation of animateness as death is the negation of the state of aliveness. Nor can something have a soul and then not have a soul and then acquire a soul again. In Odawa the concept of animateness is limitless. It can be altered by the mood of the moment, the mood of the speaker, the context, the use, the circumstances, the very cosmos of our totality. English terms imprison our understanding of our own linguistic concepts.   (King 1997: 116–17)

Though unlikely to have been instrumentally conceived as such, the acts of scholarly (re)presentation to which King refers manifest as powerful instruments of advanced colonial domination in their erasure of the very bases of Indigenous people(s)' self-definitions.

Importantly, all of this is, like the disciplinary division of knowledge, quite explicitly connected to the production of Euro-American self-knowledges as well. Proposing that "[t]he positive sciences (physics, economics, and psychology) are often seen as the crowning achievements of Western civilization," Norman Denzin notes that "in their practices it is assumed that truth can transcend opinion and personal bias" (Denzin 1996: 131). This binds the pretensions of objectivist ethnography together with conceptions of the modern Western Self, so that what is ultimately at stake is rather more than the veracity of the accounts of Europe's Others. It is indeed a serious matter, then, if it turns out that bias cannot be transcended. As King makes clear, scholarly research and writing about Indigenous peoples can be quite profoundly at odds with their own senses of themselves and their ways of being in the world. Unflagging allegiance to objectivist epistemological commitments, however, not only obscures such contradictions but makes them unthinkable—if our accounts of our Others are wrong, we lose also that portion of the account of the Self which is premised upon our highest values (including epistemological values).

In a more active sense, particular constructions of the Other enable deeply held conceptions of the Self. Drawing on the tradition of hermeneutic philosophy, Clifford notes the intentionality of the production of culture accounts, adding that "interpreters constantly construct themselves through the others they study" (Clifford 1986a: 10). This is an inherent feature of each moment of the ethnographic project, even from the decision to conduct research "in the field,"[7] and it applies as much to contemporary ethnographic research and writing as to the travelogues of the so-called Age of Discovery. The universalizing tendencies of much of Western philosophy, for instance, necessitate that Indigenous peoples be understood in particular ways. To the extent that a belief in the universality of certain ideas or practices is sincerely held, though, the resultant accounts of Others will seem entirely proper, so that the violences of misrepresentation will be due as readily to the inability to imagine other possibilities as to any deliberate obfuscation. In the founding ontologies of the various traditions of Western social theory are essential propositions about presumed universal human predispositions, such as an egoistic nature or the irresistible urge toward material acquisitiveness— propositions that, if disproved, would soon be joined on philosophy's scrap heap by the theories they once sustained. Accordingly, European social theorists as diverse as Hobbes, Locke, and Marx have imputed to "primitive" peoples such ostensibly universal human tendencies as befit their philosophies, notwithstanding that they might have flatly contradicted those peoples' accounts of themselves as borne in oral literatures and reflected in the everyday.

In this way, even when it has not been their project, Euro-American social theorists have directly participated in the violences of misrepresentation— violences that, as we shall see in later chapters, have been inexorably bound up with the colonial/advanced colonial conquest/subjugation of Indigenous peoples. Again, no malice of forethought is requisite here; a contemporary Euro-American ethnographer might believe unequivocally in the naturalness and universality of Western concepts, categories, and commitments just as surely as so many missionaries and colonial administrators once believed in the idea of "the White man's burden."[8] These are, after all, ideas and practices that have made Euro-American forms of sociopolitical organization (seem) possible. And, together with enduring popular and scholarly attachment to notions of social evolutionism—expressed in the persistence of disciplinary boundaries as much as in the idea of "progress"—these commitments have underwritten one of the central binaries of Western philosophy: civilized/ savage. Although this particular dichotomy has done much to cast the colonial conquest of the Americas as natural or necessary, it has also spoken back to its place of origin. As Thomas Biolsi argues, it has been a vital element in Euro-American self-knowledges inasmuch as "[t]he self identity or subjectivity of people in state societies . . . requires a concept of the primitive both to bound and to give content to the concept of the civilized" (Biolsi 1997: 135). As with the disciplines themselves, then, ethnographic (re)presentations of Indigenous people(s) are inseparable from dominant constructions of the Euro-American Self.

Ethical Dilemmas, Reflexivity, and Contractual Ethics

The inherent ethical dilemmas of ethnography should be fairly clear, particularly at the interstices between worlds where the inequalities of advanced colonialism are most keenly felt. But an important caveat further complicates the ethnographic project: consistent with the advices of recent postcolonial theory,[9] Indigenous people(s) are not reducible to "passive skin" awaiting the inscriptions of the colonizer. Rather, they are possessed of agency and, notwithstanding the unequal structural positions carved out for various people(s) under advanced colonialism, they resist. In particular, informants can mislead as a form of resistance. Even participant-observation is not immune to this, depending as it does on the face that the community chooses to turn to the ethnographer—this is bound to be all the more an issue in Indigenous North American communities where "anthros" are a priori suspect. Vincent Crapanzano laments that this is not always well appreciated, noting that "[a]ll too often, the ethnographer forgets that the native . . . cannot abide someone reading over his shoulder. If he does not close his book," Crapanzano continues, "he will cast his shadow over it" (Crapanzano 1986: 76). Unfortunately, resistances of this sort might not be recognized as such by the ethnographer. Likewise, the strategies of "performing" according to the scripts of the dominating society or what Bhabha calls mimicry[10] seem equally prone to going undetected and handily subverting the production of epistemologically objectivist ethnography. And, in a similar vein, Pratt draws our attention to how their engagements with the ethnographies of the colonizer render the self-representations of the colonized distinct from what is sometimes called an "authentic" voice (Pratt 1992: 7–8).

We must always bear well in mind, though, that both in our roles as academics and as members of the dominating society we are worth resisting. In each of these contexts we are, respectively, complicit in and the beneficiaries of ongoing practices of advanced colonialism. Moreover, whatever our politics, our allegiances are clear in at least one important sense: when the fieldwork is over, we return to the academy and to our privileged lives. We might think ourselves allies, but that place is much less "ours" to declare than it is "theirs" to confer. We should therefore expect to encounter resistance— what is more, we should accept it. Sadly, this advice is too seldom heeded. Much to the contrary, resistance is often read outside of the context of struggle so that it appears as mere obstinacy. Renato Rosaldo draws our attention to a conspicuous instance of this in E.E. Evans-Pritchard's (1969) celebrated work on the Nuer people of Sudan. In the personal narrative of his fieldwork experiences—where, it should be remembered, a vital component of the claim to authority resides—Evans-Pritchard recounts the reluctance of a Nuer man (Cuol) to give the name of his lineage. The ethnographer describes this as opposition and reveals a certain irritation that a simple attempt at introduction (notably, in accordance with British standards of politeness) should meet with such obduracy. But as Rosaldo points out, this assessment of Cuol's apprehension is sorely lacking in a reflexive sense of its political context:

> . . . it is measured against a norm (which probably is alien to the Nuer) of courteous conduct that requires strangers, on first meeting, to introduce themselves by giving their names. The narrator finds that the fault in this unhappy encounter lies with Nuer character, rather than with historically specific circumstances. Yet the reader should consider that, just two pages before, Evans-Pritchard has described how a government force raided a Neur camp, "took hostages, and threatened to take many more." Cuol had, not a character disorder, but good reasons for resisting inquiry and asking who wanted to know his name and the name of his lineage. (Rosaldo 1986: 91)

A reflexive sense of the political, then, gives content/reason to what might otherwise be read as intransigence. But if, like Evans-Pritchard, we express annoyance at having been denied access to that which we presume entitlement, then we commit a colonial violence of our own, asserting a right we do not have (except to the extent that we have conferred it upon ourselves).

Rosaldo also discusses the idea that records from the Spanish Inquisition can be treated as ethnographic sources—something that he finds troubling given that, far from being passive scribes, the inquisitors were engaged in the politically charged extraction of confessions (Rosaldo 1986: 79–81). The idea of the ethnographer as inquisitor is useful and revealing inasmuch as the very real power dynamics of advanced colonialism are such that the ethnographer will, at times, be very much like the inquisitor. Worse yet, the methodology of ethnographic research has a mystifying effect here—that is, while the inquisitor was more easily identifiable as a present danger to the accused, the *modus operandi* of the ethnographer is to become as one of the community, even a friend. This, however, does not enable transcendence of the power relations underlying the encounter, even if it does make them *seem* somewhat less immediate. Nor do the best of intentions fashion safety from vulnerability. As Rosaldo observes, "the fieldworker's mode of surveillance uncomfortably resembles Michel Foucault's Panopticon, the site from which the (disciplining) disciplines enjoy gazing upon (and subjecting) their subjects" (Rosaldo 1986: 92). Here is the underlying essence of participant-observation, and perhaps its greatest danger: it exploits the common human tendency to relax in the presence of the familiar. Indeed, this is often an explicit strategy where people(s) under study are especially cautious. In short, persistent presence is disarming and to the extent that informants come to trust the ethnographer among them, they might also put themselves in peril.

Once again, no specific malevolence need be involved here. Indeed, the power dynamics of advanced colonialism are such that even sincere attempts to engage with Indigenous worldviews and lifeways can have profoundly pathological effects. New Age appropriations of Lakota cultural and spiritual practices are a case in point. Feminist-inspired variants of this trend have also become increasingly prevalent—so much so that the semiotic and material "products" of their culturo-spiritual appropriations have become common-place in feminist bookstores, at feminist conferences, and in feminist periodicals (Smith 1994; Donaldson 1999). In company with elements of the Men's Movement, such groups have been at the forefront of a phenomenon that Geary Hobson has termed "whiteshamanism" (Hobson 1979). Though their behavior might be inspired by sincere reverence, too seldom understood is the epistemic dissonance that forecloses the possibility of grafting Indigenous spirituality onto lifeways otherwise rooted in and fashioned by Euro-American traditions. Consequently, the distortions they create and sustain are not well appreciated either. While Indigenous communities have reserved their most intense criticism for those who engage in the commodification and sale of their knowledges and traditions,[11] all instances of cultural and spiritual appropriation have been read by many as "a second wave of conquest, in which those who earlier seized their land may now complete the job by taking their spiritual knowledge" (Lincoln 1994: 11). When they involve a remaking of Indigenous peoples and their traditions, the violences of appropriation can be every bit as damaging as those of exclusion (Kulshyski 1997: 616), and no less the midwife of domination.[12]

Acknowledging the seriousness of this problem, John Grim wonders whether the academic study of Indigenous spirituality can be any less exploitative (Grim 1996: 359). Following Wendy Rose, he suggests that two fundamental issues inform this question: intent and the ascription of authority. Rose is clear in her conviction that members of the dominating society have as much right to study Indigenous peoples as Indigenous peoples have to study them, but it matters "how this is done and, to some extent, why it is done" (Rose 1992: 414–16). In particular, claims and accounts that deprivilege or displace Indigenous peoples' own voices are irredeemably problematic[13]— all the more so when we bear in mind that widely accepted (mis)representations have enabled colonial conquest and domination. Remembering too that the Euro-American Self has been constructed and naturalized by way of opposition to its constructed Others, we are bid to reject any lingering pretension of objectivist epistemology that might give us to believe that we need not critically situate and interrogate ourselves in/through our accounts of Indigenous people(s). Thus, Grim proposes that "[s]tudying American Indian religions entails a reflexive step that activates self-scrutiny" (1996: 359).

This demands several things of us. First, it requires that we specify clearly who we are and what has brought us to our projects. Though we are no better able to unproblematically represent ourselves, the rudiments of a reflexive account exact from us a commitment to the idea that we construct ourselves through the claims we make about Others and that an honest disclosure of our self-definitions is therefore essential if they are to make sense of what we say. That is, if our readers—and we should expect those about whom we write to be among them—hope to understand why we see other people(s) the way we do, they will need to have some sense too of who we think we are and who we would like to be. And this partially satisfies another requirement: that we contextualize, qualify, and clearly specify the limits of our authority to speak. Our voices are not legitimate substitutes for those of Indigenous people themselves—this is not to say that we cannot speak, but we must not presume to speak for or in place of our informants. Finally, and following from this last point, the deprivileging of Indigenous voices must be resisted at every turn. This calls upon us not only to take oral literatures seriously but also to cite Indigenous people themselves as legitimate author-ities, rather than participating in the conversion and appropriation of their knowledges.

Even when we abide by all of these principles, however, our political commitments can have other potentially troubling implications. Allying one's work with the counter-hegemonic resistances of one's informants, for example, is fraught with danger. Here we must take a stand on the ethical conditions of the everyday, but it is less our everyday than someone else's. Here too we wade into local politics without the possibility of requisite local competencies. And we do this from a place of conspicuous privilege, exemplified most particularly in the fact that we will be going home.[14] This is not to say, however, that we cannot or even should not take a stand—again, inaction is every bit as political a choice; it is a vote in favor of the status quo.[15] In short, whether speaking or silent, there is no avoiding the ethical dilemma. Confronted with this choice, we can elect to play the part of what Franco Basaglia calls the "negative worker" (Basaglia 1987: 154)[16] or, in Spivak's terminology, the "de-hegemonizer" (Spivak 1988c: 332). This is, in effect, the *in*organic intellectual: a privileged member of the dominating society who exploits the discursive power of that position of privilege in conscious alliance with its marginalized Others. But, crucially, we must always retain a strong sense that this can never be *our* project, lest we hazard to substitute for old tyrannies those of our own making. Indeed, even the best conceived "emancipatory" projects can have pathological effects when they presume to dictate terms of emancipation that might not be compatible with marginalized people's own articulations of the requisites of their deliverance. It is therefore all the more essential that we work to de-objectify our subjects and to privilege their voices.

What this bespeaks is an ethics of responsibility—an idea that, as we have seen, usefully reveals the point at which a viable oppositional politics can be joined with a poststructuralist project. Here ethics is conceived in terms of a right to be that is founded in the relationship between Self and Other, such that it becomes an enabling property of subjectivity (Campbell 1998: 176). To this I would add what Karena Shaw has perceptively observed about the colonial violences visited upon Indigenous peoples: "that their situation is a condition of possibility of our own, historically as well as in present times" (Shaw 2002: 59). But as its poststructuralist affinities will no doubt suggest, this should not be read as a paternalistic call to protect the interests of Indigenous people(s)—not only is it not for us to decide what those interests might be and how best to proceed, but such an orientation also has the perverse effect of re-objectifying by denying agency. Rather, an ethics of responsibility enjoins us to take responsibility also for ourselves, the ideas we advance, and the ways we propose to use them. At the same time, we must recognize that we are the Others of our Others (George 1995: 210) and that responsibility *for* the Self is simultaneously responsibility *to* the Other. In this sense, we become responsible *to* Indigenous people, not *for* them.

A frequent objection from the detractors of projects informed by post-structuralism, however, is that they disempower the ethical political stance through a descent into hyper-relativity that is inconsistent with any firm moral or ethical commitment. In more practical terms, according to this view, difficulties in concentrating resistances against some identifiable category of culpable agents hinder purposeful action. Nancy Scheper-Hughes, for example, complains that the "imagined postmodern, borderless world"—born of resisting the rigid bounding of real and imagined spaces—has occasioned a "flight from the local" and a recasting of power relations in terms so awesome that they defy any progressive response: "[o]nce the circuits of power are seen as capillary, diffuse, global, and difficult to trace to their sources, the idea of resistance becomes meaningless." Moreover, "[t]he idea of an anthropology without borders," she continues, "ignores the reality of the very real borders that confront and oppress 'our' anthropological subjects and encroach on our liberty as well" (Scheper-Hughes 1995: 417).

Scheper-Hughes is quite right in reminding us that the material exigencies of power are no less real for their contingency. But we might press her to show us where this has ever been seriously questioned by those to whom she ascribes the (here pejorative) label "postmodern" and how, in light of this, they have affirmed the impossibility of a politics. Acknowledging the capillary nature of power does not entail paralysis. Nor does it deny the material expressions and experiences of power as embodied in borders or barbed wire and inscribed (not least in blood) on bodies. Rather, it demands of us that we confront our own less visible complicities in these articulations. True enough, such a view of power does not allow us the false comfort of laying blame for all wrongs and suffering at the feet of some small number of identifiable tyrants; we are quickly disabused of any sense that power might be so easily and unproblematically managed by the politically enlightened. And as already noted in reference to the origins of disciplinarity and the violences of (mis)representation, instrumental design is not always found at the root of every instance of oppression. But far from disempowering action, this is an enabling move inasmuch as it underscores the folly of the all-too-common fallacy that structural arrangements whose very founding is oppressive—as with disciplinarity, for example—can somehow be made benign simply by replacing some or all of the people who populate them: an ill-conceived hope that leaves macro inequalities and all of their pathological effects well intact while substituting the illusion of emancipatory change for a real politics of resistance.

Our ability to adopt a politicized ethics of responsibility does, nevertheless, seem sorely challenged when we venture to represent. First, we lack the cultural competencies to write about Indigenous peoples absent the violences of misrepresentation. Participant-observation is therefore prescribed as the means by which to develop such competencies. As we have seen, however, cognitive limits and the unequal power relations between ethnographer and informant(s) put the efficacy of this research strategy in serious question. Moreover, it is ethically problematic in itself to the extent that it enables researchers to exploit the trust they develop with their informants over time, extracting testimony from them that they might not otherwise wish to divulge—it is important to remember that what is shared with the ethnographer is never shared in confidence (Homan 1992: 22–23). Again, the inability to separate our work from our own privileged places in sociopolitical time and space makes it prone to advanced colonial complicities and the commission of other violences. How, then, do we implement our ethics commitments?

The unsatisfactory answer has come in the form of contractual Ethics. This style of Ethics may actually have precious little to do with ethics.[17] On the face of it, it seems quite well intentioned, legislating, in effect, the terms of the encounter between researcher and informant. Full disclosure of research aims is mandated, as is the promise that informants can disengage at any time should they become ill at ease. In short, informed consent is solicited and, if granted, a contractual agreement is concluded—sometimes in written form, other times verbally—whereby informants are assured of the inviolability of the terms to which they have agreed. For good measure, the whole affair is subjected to an Ethics review process at many universities so as to ensure that researchers are conforming to accepted standards of Ethical conduct. But contractual Ethics is rather more complicated in practice than it might appear on first gloss. Ironically, it is itself the source of some especially thorny ethical problems. Moreover, like disciplinarity and ethnographic (re)presentation, it has its own complicities in the violences of advanced colonialism.

Of most immediate concern are the problems unique to this project. The reader is invited to imagine the effect of my arrival in Lakota country armed with contractual Ethics guidelines: a White male member of the dominating society coming from the East with papers to be signed—papers full of promises that the rights of the signatories will be guaranteed in perpetuity. This might be comical were it not so insensitive to the long history of broken treaties that were to have safeguarded their Indigenous signatories from the predations of ever-expanding European colonies—the original papers full of promises. Enter now the academic with intellectual treaties in hand—more papers full of promises. Far from imparting any assurance of goodwill, resort to contractual Ethics threatens, in this context, to insult local sensibilities. Contracts do not enjoy any place of reverence for many Indigenous people.[18]

Equally problematic is the pretension to dictate what constitutes ethical conduct in Indigenous communities. If our representations of Indigenous people(s) are compromised by the distortive interventions of decidedly Western concepts, categories, and value commitments, we can scarcely hope to pronounce upon local standards of ethical conduct without perpetrating similar violences. And this is something that is already well understood in Indigenous communities, as Cecil King tells us:

> We acknowledge, with gratitude, the attempts by the National Endowment for the Humanities and the American Anthropological Association to regulate researchers by guidelines or codes of ethics. However, for most of us, these efforts are part of the problem. For we must ask: Whose ethics? In this era of aboriginal self-government, it is not for the outsider to set the rules of conduct on our lands and in our communities. It is our right and responsibility as aboriginal nations to do that. It is the right and responsibility of researchers to respect and comply with our standards. The dictates of Western science and the standards of behavior enshrined by associations of researchers dedicated to the advancement of social science may or may not be compatible with the code of ethics of our aboriginal communities. (King 1997: 118)

The intellectual treaties of contractual Ethics thus follow those other treaties from the colonial encounter in codifying aims and objectives that are principally those of the dominating society. What could more clearly signal to people in Indigenous communities that they should not risk talking to us?

All of this is compounded by a range of practical problems with dire ethical implications. Assuming that we have secured informed consent from one person, for instance, what about the consent of others who might in some way be implicated in what that person has to say? This is an even bigger problem in communities with a more developed sense of collectively held knowledges; the consent process under contractual Ethics guidelines is highly individualized, reflecting the liberal commitments of the dominating society, but not necessarily fitting as well with those of many Indigenous communities (Smith 1999: 118). Even more problematic, we risk putting people in jeopardy in the very moment of soliciting their informed consent—particularly in politically divided communities. Just in approaching someone, we can have the effect of forcing them to declare a political position when their personal survival strategy might have been to cultivate an air of neutrality; or, we might force someone to declare against their conscience—choosing to participate in our project, or not—for fear of consequences. We are thus unable to escape being caught up in local power dynamics and, in some cases, might even unwittingly ally ourselves with the purposes of locally dominant actors.

As students of International Relations undertake more and more ethnographically based projects, they are becoming answerable to contractual Ethics review processes. But the suspiciously minded—and I am one—may find ample room for cynicism in the idea that ad hoc contractual arrangements are a sufficient substitute for critical engagement with ethnographic research methods and the ethical issues that attend them. Such instruments might well limit liability, but they are no assurance of the limiting of violence(s)— much to the contrary, they have serious pathological effects of their own. We would do well, then, to take seriously the possibility that Ethics review processes, whatever their founding intentions, can in practice be more protective of the ethnographer than her/his informants. Indeed, Roger Homan worries that "Ethical codes, like other legalistic formulations, tend to set out procedures rather than stress the values which they are designed to safeguard," with the result that "[t]hey invite observance in the letter rather than in the principle" (Homan 1992: 325). Proposals submitted to Ethics review processes may thus pass scrutiny despite specifying only procedures unaccompanied by any comprehensive declaration or elaboration of the particular ethical commitments that underlie them. This threatens to reduce the consideration of ethical issues to a quasi-legalistic exercise that, as Homan warns us, can have the unintended effect of allaying some researchers' sense of responsibility for the risks to their informants once a release has been signed: "codes . . . close down rather than sustain the moral responsibilities of the individual researcher" (Homan 1992: 330). Contractual Ethics thus does nothing to advance an ethics of responsibility, making codified standards, and not one's informants, the primary referent of obligation.[19]

Conversation

Ethnography, ethics, and disciplinary International Relations are intimately bound together, not least around the matter of conversation. As I have argued, even where extradisciplinary forays have occasionally been made by those working in International Relations, the tendency to appropriate as opposed to establishing conversations has left disciplinary boundaries and their advanced colonial complicities well intact. Similarly, the practice of appropriating Indigenous people(s)' knowledges rather than privileging their voices so that they might be engaged in conversation only deepens their subordinate position by reconfirming the authority with which the ethnographer speaks and writes. And just as we should not expect the varied pathologies of disciplinarity to be resolved through ad hoc interdisciplinary borrowings, those of ethnography will continue unabated so long as its subjects are objectified and held at bay.

This is not a call, however, to bring Indigenous people(s)' voices into International Relations, as this would make the paternalistic mistake of presuming a responsibility *for* them rather than a responsibility *to* them. The former stance suggests that we think their knowledges our wards, with little value independent of our interventions; the latter that we think them the equals of our own. Thus, appropriation of Indigenous knowledges in ways that make them fit harmoniously with the dominating society's predetermined concepts and categories is also an exercise in their devaluation—in fitting in, after all, they will only have confirmed what was already "known." Such appropriations may thus be counted among the practices of advanced colonialism. However, if voiced on their own terms in ways that do not always sit well with our own commitments, then, as Deloria argues, something of the inherent value of Indigenous knowledges may be revealed to us (Deloria 1997: 220–21). Indigenous people(s) do have things to say about the subject matters we routinely engage, and this calls upon us to reflect critically upon the reasons why their voices have not been heard and what we might do differently in this regard.

In the end, though, it might be that the most important things we can learn about international relations are actually things we have yet to learn about ourselves. It might also turn out that this promise is conditional upon the extent to which we are able to engage disciplinary International Relations' Others—and the historical contexts of disciplinarity are such that they *are* International Relations' Others as surely as they are Europe's Others—in a conversation between authoritative equals. This is why it is not enough that other voices simply be "brought in" to the discipline—the inevitable outcome of such a strategy is that they will languish in a position of lower hierarchical order in much the same way that extradisciplinary conceptual borrowings do. A conversation, on the other hand, is about speaking *with* others as opposed to speaking *about* or even *to* them. It presumes not only that others have something to say that is worth hearing, but also that they deserve a response. In short, conversation, involving both reciprocity and a mutual authority to speak, presumes relative equals. Of course, the material realities of advanced colonialism sustain sites of privilege and margin such that we must always take care not to confuse these commitments with a more liberal notion of equality and venture to imagine that the authority to speak is emancipation in itself. What conversation does do, however, is to broaden the intertextuality of the hermeneutic circle, thereby effecting an interruption of the hegemonologue. As a vital aspect of our responsibility to our Others it also gives essential content to an ethics of responsibility.

Geertz proposes that Anthropology needs to develop a new ethnographic project that focuses not just on "natives" but on anthropologists as well. And viewing the disciplines—Anthropology among them—as "ways of being in the world," he urges us to consider that "[t]hose roles we think to occupy turn out to be minds we find ourselves to have" (Geertz 1983: 155). Likewise, when we read a text, he calls on us to situate and contextualize the author in it: "it still very much matters who speaks" (Geertz 1988: 7). Geertz's hope is that, if we heed these advices, the inevitable violences of subjective interpretation might be somewhat alleviated. While our inability to free ourselves from the fetters of advanced colonialism necessarily limits us in this regard, it might not be too optimistic to hope that we may at least become more alert to the intertextual nature of the seemingly singular text and find in this the basis for conversation. What this means is that our aim must be to make it possible for Indigenous people(s) to speak authoritatively about international relations instead of presuming to do the speaking for them.

Having said this, I do seem to speak for and about Indigenous people(s) in much of what follows. This has been necessary to the extent that International Relations has yet to widely embrace the sort of methodologically intertextual approach that would allow us to engage directly with Indigenous knowledges about the discipline's core subject matters. Thus, if I speak about Indigenous people(s) it is in that same spirit in which Fabian hopes we might "transform ethnography into a praxis capable of making the Other present (rather than making representations predicated on the Other's absence)" (Fabian 1990: 771). In this sense, the present project is still only a prelude to a conversation and bears no pretensions as to the authority of its author's voice. Accordingly, none of what follows should rightly be regarded as anything more authoritative than a considered set of interpretations. Of course, there are good and well-established grounds upon which to argue that this should always be the case when one approaches a subject matter with which they do not share a thorough and intimate lived experience. Nevertheless, the point is one that it is particularly important to underscore in this instance, given the long history of spurious accounts of Indigenous people(s) that have issued from ostensible "authorities" and the nefarious political purposes to which they have sometimes been turned—often buttressed by a claim to authenticity rooted in the personal narratives of participant-observation.

If moral dilemmas have been my steadfast companions throughout this project, then, I have also learned to befriend them. They have served constantly to remind me of the theoretical commitments from which I set forth and to urge my conscience to the fore as an arbiter of whatever claims, however well qualified, I might venture to make. And they have simultaneously brought my methodological commitments into sharp relief. I have attempted to answer the politico-ethical dilemmas of implementing open research methods by approaching and citing only persons who are already openly politically active. (Notwithstanding that this project can be no more than a prelude to conversation, conversation remains its central methodological commitment inasmuch as I endeavor to privilege Indigenous people(s)' own knowledges and texts and to situate them vis-à-vis those of their ethnographers and those of International Relations.)The approach taken, then, is decidedly intertextual. Recognizing that the parceling off of knowledges and method-ologies risks not only tunnel vision but socially irresponsible praxis as well (Nissani 1997: 209), I also eschew disciplinary allegiances and seek wherever possible to destabilize the secure foundations of disciplinarity itself. And to the extent that these commitments also enable revelations about the complic-ities of disciplinary International Relations in advanced colonial domination over Indigenous people(s), I am self-consciously deploying science *in* history (Clifford 1988b: 120) as read through a postcolonial lens.)

Notes

1. See Lagrand (1997).
2. The at least mildly pejorative term "anthro"—a common appellation used in many Indigenous communities to describe those whom they regard as usurping their authority to secure their own definitions of who they are—underscores this mistrust.
3. See Austin-Broos (1998).
4. On what she calls "the cultural politics of ownership," see Whitt (1995a).
5. As Randall McGuire acknowledges, the realities of advanced colonialism are such that this turns out to be more than just pretension: "In the United States, archae-ologists and anthropologists have been the authorities on Native American pasts, and this authority has given us a power over those pasts. Courts of law and government commissions call us as expert witnesses and have often given our testimony more weight than that of tribal elders" (McGuire 1997: 65).
6. See chapter 6 of this book on the violences caused by the application to Indigenous contexts of social theory committed to a Hobbesian-inspired account of human nature.
7. Johannes Fabian argues that "[t]he need to go *there* (to exotic places, be they far away or around the corner) is really our desire to be *here* (to find or defend our position in the world)" (Fabian 1990: 756; emphasis in original).
8. On Anthropology's complicities in the advancement of U.S. national and strategic interests, see Price (1998) and Reddock (1998).
9. In particular, I am thinking here of the ideas of hybridity and mimicry associated with Bhabha's work.
10. See his essay, "Of Mimicry and Man: The Ambivalence of Colonial Discourse" in Bhabha (1994).
11. See, e.g., Jocks (1996) and Rose (1992).

12. This is not to suggest that Indigenous cultures are, should, or can be static. The issue is not with culture change, but with how, why, and by whom it is brought about, as well as with the silencing of resistance that follows from unequal speaking positions.

13. In addition to sharing accounts of similar experiences of her own, Rose tells of an instance in which a Cree woman (Sharon Venne) who attempted to challenge a leading figure in pop-culture feminist circles (Lynn V. Andrews) on the inconsistencies in her accounts of Indigenous cultural and spiritual knowledges was shouted down by Euro-American women and chastised for "not knowing the 'inner meaning' of native culture as well as Andrews" (Rose 1992: 414–15). See also Cook-Lynn (1993).

14. Indeed, the very fact of our being there bespeaks a position of privilege. Not everybody enjoys the luxury of mobility that makes fieldwork possible. The ability to travel limits who is able to produce ethnographies of whom. See Natrajan and Parameswaran (1997).

15. See Fabian (1990: 767–68) and Whitworth (2001: 158–59). It is in this sense that Philippe Bourgois argues for a fuller conception of ethics that would impel us to take a moral stand on the issues confronting our informants (Bourgois 1990).

16. Basaglia borrows this specific phrase from René Lorau, but it is consistent with his own idea of radical "negation" by privileged members of society (intellectuals in particular) as a means to destabilize the secure foundations of others' marginalization. The "negative" worker is thus the functional opposite of what Paul Baran (1961) calls the "intellect worker" (the organic intellectual of the status quo).

17. I use "Ethics"—as distinct from "ethics"—to signify the codification of standards of conduct in research involving human participants.

18. According to Gloria Dyc: "As the Lakota witnessed the breaking of treaties during westward expansion, for example, they came to believe that the signing of written documents was no real protection against people who were bent on appropriating Indian lands and resources. They doubted the integrity of the Europeans and questioned their dependency on the written word" (Dyc 1994: 215).

19. Contractual Ethics might thus be read as a form of self-delusion in some instances; a salve for the conscience of the researcher. On self-delusion, denial, and ethnography see Fine (1993).

CHAPTER 4

Lakota Lifeways: Continuity and Change in a Colonial Encounter

I begin this chapter knowing that I am about to violate at least the spirit of the commitments I have laid out for myself in the preceding chapters. Still working through a pre-conversational moment, I find that I am obliged to engage in at least some small measure of (re)presentation in making the case for real engagement with Indigenous people's voices. With the aim of minimizing the violences of this exercise, however, I have chosen not to rely on participant-observation in support of my claims, placing the primary emphasis instead on autoethnographic accounts and narratives—that is, Indigenous people's own texts about themselves. Pratt uses the term "autoethnography" to indicate a form of self-representation by colonized subjects that, because it engages with the ethnographic texts of the colonizer, is distinct from what is sometimes called the "authentic" voice (Pratt 1992: 7–8). But, reflecting the circumstance that the autoethnographic voices of interest here are somewhat more complicated by virtue of their connection to oral literatures, my usage falls somewhere between the two. While each of the autoethnographies drawn upon herein must certainly have been influenced by the exigencies of life in what Pratt calls the "contact zone" of the colonial encounter, the more communal nature of oral literatures suggests that textual revisionism is likely to move more glacially and less idiosyncratically than might be the case where single-author-ized written forms are concerned. In any event, the autoethnographic voices consulted in the writing of this chapter might be described as "authentic" in the important sense that they are quite widely recognized as authoritative within their own ethno-cultural context.

Following from this, a note regarding selection of autoethnographies is also in order. This matter is immediately complicated by practical disjunctures between oral literatures and the Western conventions by which texts are author-ized. In outlining an account of pre-reservation Lakota cosmological commitments and lifeways, for example, those whose accounts I have chosen to draw upon include Nicholas Black Elk, Luther Standing Bear, and Thomas Tyon. All three reached adulthood as members of pre-reservation Oglala[1] Lakota bands and should thus be listened to at least as fully as twentieth-century anthropologists. As a people with an oral literary tradition, the Lakota have no written texts predating contact. Standing Bear, however, received a Western education and authored his own books and therefore is not as easily dismissed in deference to orthodox academic sensibilities as, lamentably, contemporary bearers of oral history might be. Tyon, whose narratives (published by physician and amateur ethnographer James R. Walker[2]) draw on consultations he conducted with his own elders and other Lakotas around the end of the nineteenth century, is also their author.[3] The case of Black Elk is somewhat more complicated: each in their turn, John G. Neihardt (1979) and Joseph Epes Brown (1953) published accounts of their interviews with the celebrated Oglala *wicasa wakan* (usually translated as "holy man"), but these bear heavy Christian overtones that have been the source of some considerable controversy.[4] For this reason, I have drawn only on Raymond DeMallie's *The Sixth Grandfather* (1984) which, working from the original shorthand notes taken during the Black Elk interviews rather than the transcripts typed afterwards, exposes many of the textual liberties apparently taken by Neihardt and Brown in their published versions. In their narratives published by Walker and DeMallie, we are thus offered the chance to hear the voices of Tyon and Black Elk respectively.[5]

Of course, none of this should be taken to suggest that these or any other texts are ever free of the multifarious workings of domination and resistance that characterize the colonial encounter and its legacies. Hybridizations, whether by coercion or willing adoption/adaptation, are always a central feature of life in the contact zone.[6] It is worth noting, for example, that even absent the interventions of Neihardt and Brown, we cannot cleanly separate out the influences of the colonial encounter from the words of Black Elk who, in his later life, became a Catholic catechist.[7] Nor are the accounts by Standing Bear and Tyon any more likely to correspond in every particular to what might be called a "pristine" aboriginal condition. If nothing else, the vagaries of translation from the Lakota language will have introduced distortions wherever English lacks the phraseology to express a uniquely Lakota concept or idea.[8] Even texts preserved by Lakota people in their own language do not speak unproblematically to a reader separated by time and/or space from the cultural competencies requisite for an appreciation of the subtleties of nuance and context.[9] And this dilemma is no less acute when we come to the more recent (including contemporary) autoethnographical voices cited later. Notwithstanding such limitations, however, each of these narratives allows us at least a glimpse of ways of knowing and being in the world that, in many vital respects, are radically different from those of the dominating society. These autoethnographical voices, then, are indispensable not only to the (re)presentations I venture to make in this chapter, but also for their counter-hegemonic potential to unsettle the hegemonologue.

Autoethnographies of the Hoop: Lakota Cosmology and Traditional Lifeways

Even as we acknowledge the transformative potential of Indigenous autoethnographies, we should be wary of the very serious implications of tearing this or that aspect of a given people's lived experience and worldview from its proper context and subjecting it to the deforming constraints and impositions of foreign ontological and epistemological commitments.[10] In particular, serious consequences may arise from attempts to make literal translations of core Indigenous concepts. In an influential ethnography of the Yanomami people of Amazonia, for example, the word *waitheri* is translated as "fierce" (Chagnon 1988: 987). This rendering of *waitheri*, a sophisticated concept denoting conduct regarded by the Yanomami as virtuous, seems to confirm accounts that cast their in-group lifeways as inherently combative and sustained by coercion. Jacques Lizot, however, disputes both the translation and its implications, countering that the word is not simply descriptive of a state of being for the Yanomami, but signifies a highly nuanced concept with a broad spectrum of meaning that includes, simultaneously, courage, gallantry, recklessness, and stoicism (Lizot 1994: 857). "Fierce," according to Lizot, "occurs only at the far edge of the spectrum of possible meanings, to describe an extreme behavior" (Lizot 1994: 857). He continues:

> To be waitheri is to be courageous and stoic, to have no fear of others' aggression, to refuse submission, to be capable of opposing the will of others, and to stand up to them; it is also to be able to endure the greatest physical or psychological suffering. It is interesting to note that the two animals which, in the eyes of the Indians, best embody waitheri behavior are the coati, for its bravery, and the sloth, because it "does not die"— in other words, because it endures the most excruciating wounds; it is

difficult to kill. It is not a killer animal like the jaguar that has been chosen to represent the waitheri ideal, but rather two other animals and, if these animals are different, it is because the Indians recognize at least implicitly the two poles of the semantic field covered: courage and stoicism. . . . Perhaps it is clearer now why intimidation cannot constitute the basis of Yanomami social relations: submission is contrary to [Yanomami] morality; it is dishonorable.   (Lizot 1994: 857)

Seen in this light, the choice of "fierce" as the sole translation for *waitheri* seems a serious distortion not only of the idea itself but also of the comportment and lifeways of the Yanomami people more generally. In this sense too, it stands as a powerful warning: in attempting to learn about and from the worldviews and lifeways of Indigenous peoples, it is crucial that we avoid the mistake of constructing that which may seem nominally familiar in terms of what we might imagine to be correlates in our own lived experience. In short, we must be ever mindful of the prismic mystifications of intersubjectivity and endeavor (as best we can) to take seriously the voices, ideas, and perspectives we encounter on their own terms and in their appropriate cosmological contexts.

We are confronted with this challenge immediately upon beginning to consider traditional Lakota cosmology, wherein existence is expressed as a circle rather than in the linear terms of Western cosmology. As Tyon tells us:

The Oglala believe the circle to be sacred because the Great Spirit caused everything in nature to be round except stone. Stone is the implement of destruction. The sun and the sky, the earth and the moon are round like a shield, though the sky is deep like a bowl. Everything that breathes is round like the body of a man. Everything that grows from the ground is round like the stem of a tree. Since the Great Spirit has caused everything to be round mankind should look upon the circle as sacred for it is the symbol of all things in nature except stone. It is also the symbol of the circle that marks the edge of the world and therefore of the four winds that travel there. Consequently, it is also the symbol of a year. The day, the night, and the moon go in a circle above the sky. Therefore the circle is a symbol of these divisions of time and hence the symbol of all time. (Quoted in Walker 1917: 160)

This account by Tyon bears unmistakable overtones of the Lakota sense of the intrinsic relatedness of all things. Moreover, as Black Elk explains, the power that sustains life flows directly from one's connection to this circle of relatedness, a connection that is upheld, in part, via literal expressions of the circle in everyday life: "You will notice that everything the Indian does is in a circle. Everything they do is the power from the sacred hoop . . . The power won't work in anything but circles" (quoted in DeMallie 1984: 290–91). Tyon concurs in this, citing it as the reason why the Lakota lived in round tipis that they arranged in a circle (quoted in Walker 1917: 160). And just as the power and unity inherent in the circle is important to the well-being of individuals, so too is it crucial to the health of the nation. The sacred hoop of the nation is a metaphor, derived from the camp circle, for the holistic unity of the Lakota people. Like the tipis that make up the camp circle, the nation is seen in terms of a hoop wherein no one constituent part is logically or implicitly prior to any other and such that all are equally necessary to complete the unity of the circle. The significance of the circle, then, is rooted in the assumption of an essential continuity from individual, through nation, to all elements of the cosmos, and back again. In fact, no one of these can be separated out from the others, since together they constitute a single totality encompassing all of Creation.

This resists Western cosmology's hierarchical siting of humans apart from and above nature. As Thomas J. Hoffman suggests, the European view of nature as adversary and/or subordinate can be traced to the very bases upon which the claim to know is founded, exemplified most particularly in Plato's privileging of the rational over the experiential. This crucial first move in the hierarchical separation of ways of knowing and being in the world is reflected also in St. Paul's rendering of the soul as superior to the body—itself decisive to the forging of a tradition that "emphasizes either transcending the body, the physical, or splitting from it" with the result that "[t]he spiritual becomes remote from the day to day" (Hoffman 1997: 453). The Judeo-Christian Creation story—in the course of which, it is worth noting, Adam and Eve are advised by God of their dominion over all other worldly beings and instructed to subdue the earth[11]—is punctuated with perhaps the most profound disparagement of the natural world: following original sin, humanity's enduring punishment is to be cast out of paradise into a nature far less hospitable than what they knew in Eden. The Euro-American view of nature thus originates not only from a belief in human superiority but owes also to the idea of a better world both lost and yet to come. The spiritual and the mundane are thus parceled off into discrete realms, the natural world being situated entirely in the latter and constructed as simultaneously the adversary and material birthright of humanity; from this perspective, it might also be regarded as the prototypical Other.

Lakota cosmology gives rise to epistemological commitments that are very different from Plato's in that they resist the hierarchical separation of knowledge realms and ways of knowing. Here rational thought is but one aspect of a much more comprehensive approach to knowing that also includes such processes as intuition and dreaming (Bunge 1984: 75; Irwin 1994). This eschews the linearity of the rational knowledge/experiential knowledge binary in deference to the organic holism of the hoop. And this defies the construction of hierarchy—as Paula Gunn Allen explains, the idea of the sacred hoop "requires all 'points' that make up the sphere of being to have a significant identity and function, while the linear model assumes that some 'points' are more significant than others" (Allen 1992: 59). Knowledge and knowing, then, are themselves organically constituted through and situated in the organic totality of the cosmos. Here "spiritual" and "mundane" are meaningless—indeed nonsensical—signifiers since spirituality cannot be separated out from other forms or bases of knowledge; nor can it be assigned some separate realm of its own. There is thus no basis upon which to set humans or human consciousness apart from the natural world. Nature, therefore, is not something that must be overcome, with the result that the accent is on harmony over struggle; as Robert Bunge stresses, emphasis is placed on adjusting to nature, not subduing it (Bunge 1984: 94). In keeping with the endless unity of the circle, all things and beings in the universe exist together in a balance fixed long ago (DeMallie 1987: 31). Adjusting to this balance ensures its maintenance and, by extension, the security of all in Creation.[12] Unlike the Judeo-Christian heritage, the Lakota were never cast out of their Eden; much to the contrary, they are inseparable from it.

The assumption of the fundamental interrelatedness of all things and beings is expressed in the Lakota maxim, *mitakuye oyasin*—usually, if somewhat imperfectly, translated as "all are my relatives" or "we are all related." *Mitakuye oyasin* is in no way regarded as a normative proposition, but as a statement of simple fact whose falsity is so completely unthinkable that it may rightly be regarded as an aspect of traditional Lakota "common sense." According to Fritz Detwiler, from this perspective, simply by virtue of their being part of the sacred hoop of the cosmos, "all beings are related in a way that reflects the ontological oneness of creation" (Detwiler 1992: 238). As Detwiler explains it:

> The Oglala understand that all beings and spirits are persons in the fullest sense of that term: they share inherent worth, integrity, sentience, conscience, power, will, voice, and especially the ability to enter into relationships. Humans, or "two-leggeds" are only one type of person. Humans share their world with Wakan and non-human persons, including human persons, stone persons, four-legged persons, winged-persons, crawling persons, standing persons (plants and trees), fish-persons, among others. These persons have both ontological and moral significance. The category

person applies to anything that has being, and who is therefore capable of relating.   (Detwiler 1992: 239))

From this perspective, given the emphasis on adjusting to—as opposed to subduing—nature, and inasmuch as other peoples are, like the Lakota themselves, related parts of a supremely holistic cosmos, bringing ruin upon them in warfare or by any other means would be inconsistent with Lakota cosmological commitments. Moreover, it would be self-destructive since it would fragment the sacred hoop upon which all life depends—a hoop must be complete to be a hoop at all. *Mitakuye oyasin*, then, expresses not only the interrelatedness, but also the interdependence of all elements of Creation.

Linda Tuhiwai Smith identifies respect as a vital prerequisite to maintaining cosmological balance (Smith 1999: 120). Similarly, Laurie Anne Whitt argues that "as an ethical and cognitive virtue" in many Indigenous societies, wherein it "mediates not only human, but human/nonhuman relationships," the notion of respect operates such that, "since everyone and everything has important functions, they deserve to be respected for what and how they are" (Whitt 1995b: 243). And this outlook derives, in no insignificant way, from an assumption of both epistemic and cosmological diversity. Accordingly, as Howard Harrod explains:

> Even though there were religious interchanges among groups, Native American peoples were not motivated to convert others, because they did not believe that one religion was true while the other was less true or even false. Evangelism and conversion were not the point of these religions. Indeed, to offer the power of one's central religious rituals to another was viewed as dangerous since such activity might cause a diminished relation of one's group to life-giving powers.   (Harrod 1995: 103–04)

Similarly, according to Vine Deloria, Jr.:

> No demand existed . . . for the people to go into the world and inform or instruct other people in the rituals and beliefs of the tribe. The people were supposed to follow their own teachings and assume that other people would follow their teachings. These instructions were rigorously followed and consequently there was never an instance of a tribe making war on another tribe because of religious differences.   (Deloria 1992: 36)

All of this is borne out in Lakota oral history, wherein no account will be found of any attempt or inclination to convert others, past or present.[13] It should also be noted that if, as Deloria maintains, no wars were fought over "religious differences," this would almost certainly mean that divergent lifeways would not have been a source of derision either given that, as with most Indigenous peoples, spirituality for the Lakota was not ontologically separable from any other aspect of life or existence, however seemingly mundane to Western sensibilities. In this regard, the absence in most Indigenous languages of any pre-contact word by which to indicate "religion" or "spirituality" as discrete spheres is particularly telling (Kasee 1995: 84).

A most significant aspect of the assumption of unproblematic epistemic and cosmological diversity is that it turns on the idea that seemingly incompatible truth claims are not necessarily mutually exclusive. Nonlinear conceptions of time, space, and existence do not lend well to the construction of hierarchies, meaning that there is no need for one "truth" to be assigned a higher or lower order of rank than any other. Put another way, what is true for one person need not preclude a different truth for another. More importantly, the sanctity of one's own truths is in no way contingent upon remaking Others so that they accept and conform to them. As Sitting Bull said in 1876, "It is not necessary for eagles to be crows." Much to the contrary, this would endanger the sacred hoop whose continuity is ensured by the maintenance of unity and balance between all of its constituent parts, between all that is. And this, in combination with the operant notion of respect, resists denigrating practices of Othering by, in effect, valorizing difference.[14] Similarly, the derivation of ontological significance from a relational dynamic obviates any recourse to some arbitrary evaluative criteria by which the Other's inherent worth might be assessed—ontological and moral significance inheres automatically in the state of being that makes it possible to relate. It is thus that other peoples (including, but not limited to, human peoples) are naturally deserving of respect; it is thus that an ethics of responsibility is operationalized in the everyday.

This broad outlook, which wrought no impetus to enforce conformity of others to one's own will or ways, was also reflected in the political structures of decision-making authority characteristic of a Lakota *tiyospaye* (band). A *tiyospaye*'s council was called to convene whenever (and only as) needed to fulfill its collective decision-making function. Although membership was extended by formal invitation, all members of the *tiyospaye* were free to speak in council. Consistent in some ways with the principles of ancient Athenian democracy,[15] it was required that all decisions be products of consensus rather than majority vote—so long as consensus could not be reached on a given question or issue, no decision could be rendered. This was fundamental to the Lakota conception of authority expressed as *Oyate ta woecun*, translated by Standing Bear as "Done by the people" or "The decision of the Nation"

(Standing Bear 1978: 129). Even this, however, could not be made binding upon others.[16] Although a form of executive authority did come to prevail in matters of immediate urgency—such as when the *tiyospaye* was under attack—it was completely specific to and coterminous with the special conditions that called it into being in the first place. The *akicita*, for example, performed a nominal and transitory policing function during buffalo hunts and were invested with considerable powers of censure in ensuring that the hunt remained a coordinated effort and that no individual did anything that might jeopardize its success. Still, even in this temporary form, authority was not automatically vested in any one designated individual or group, but was deferred to those most adept at dealing with the particular concern at hand.[17]

Even so, the Lakota were not without identifiable leaders, though we are challenged here to resist imposing the meaning of the more common rendering of the "chief." Individual bands were guided by *itancan*—patriarchs who could attain their positions only by way of positive attributes of character, earning them the respect and admiration of the *tiyospaye*. Though they held a place of honor in council, the *itancan* were not possessed of any independent decision-making authority that could be made binding upon their bands or any individual members thereof. This fact was reflected in the repeated frustration of efforts by colonial administrators to elevate the authority of the *itancan*. As no agency of the United States was capable of interacting directly with a whole people and given that emissaries from Washington lacked the patience to await the often lengthy deliberations of the councils, attempts were made to have respected members of the bands negotiate treaties and other agreements on behalf of their people. But, as Matthew Hannah has observed, even into the reservation era U.S. officials "erred in believing, first, that Oglala chiefs commanded unquestioning assent among their people, and second, that certificates, trips to Washington, or any other such blandishments could increase what authority the chiefs did have" (Hannah 1993: 424).[18] To the extent, then, that they could ever presume to speak on behalf of their *tiyospaye*, it would have to be on matters where collective decisions had already been reached in council.[19] Moreover, the status of the *itancan*, contingent as it was on the reverence of their community, could evaporate quickly should they attempt to exceed their authority or otherwise fail to adhere to high standards of character (Powers 1975: 202–03).

None of this is to say that the pre-reservation Lakota were a people without conflict. Certainly, band councils had a need to resolve persistent deadlocks on important issues in respect of which consensus could not be reached. Dissenters in such cases would be subject to the social discipline of peer and family pressure (Price 1994: 451). Such exhortation was grounded in an ethics of responsibility, expressed in the social expectation that individuals comport themselves in a manner consistent with "buffalo virtues," placing communal interests above individual ones after the manner of buffalo bulls that would instinctively sacrifice themselves in defense of the herd (Rice 1991: 126). In the event that this too failed to break the impasse, resolution eventually came through the mechanism of secession—dissenters might join another existing *tiyospaye* or, if their numbers were sufficient, found a new one (Price 1994: 451–52; Lonowski 1994: 154). As all bands remained part of the Lakota *Oyate* (people or nation) and would still come together in the summer months to perform ceremonies and to take part in the communal buffalo hunts, secession was an accepted and legitimate mode of dispute resolution that implied no lasting enmity. Still, being a more drastic solution, it also served to deter intransigence by the majority and, simultaneously, prevented exercises of tyranny by preponderance of numbers. As in the view of the relation of humans to nature, the political sphere was thus constituted in deference to the cosmological emphasis on adjustment of human conduct so as to maintain balance.

Likewise, in the inter-national realm, conflict with other peoples involved efforts to restore balance perceived to have been temporarily lost. In this context, the existence of the revenge complex[20] is particularly significant inasmuch as it both sustained low-intensity violence between groups and mitigated against disproportionate acts of retaliation—an exercise mandated by and subordinated to the imperative of maintaining/restoring balance.[21] Here too, then, the point is not to deny that conflict was part of the aboriginal condition but, rather, is to highlight the absence of a general anarchy. Simply put, functional non-state mechanisms worked to furnish political order. What would have been unthinkable as a persistent feature of life on the pre-Columbian Northern Plains is large-scale exterminative warfare; such conduct would have been seen to jeopardize one's own well-being by threatening to break the sacred hoop.[22]

Ethnocide and the Attack on the Tiyospaye

Exterminative warfare came to the Northern Plains with the growing presence of Euro-Americans in the mid-nineteenth century, intensifying around the eventual campaigns to confine the Lakota and neighboring peoples to newly established reservations. The pre-reservation Lakota were the real-life embodiment and quintessence of the archetypal "Plains Indian," distorted images of whom have long been a feature of Euro-American popular culture.[23] That the customs and folkways of the Lakota should have so strongly influenced the homogenized popular image of "the Indian" in general[24] is likely the result of having been among the last of the Indigenous peoples on the territory claimed by the United States to be "subdued." The infamous 1890 massacre at Wounded Knee Creek, in which nearly 300 Lakotas of Spotted Elk's Mnikowoju band were killed by the U.S. Army's 7th Cavalry, is generally regarded as having marked the final "taming" of the American West.[25] At any rate, the Lakota had already figured prominently in the collective imagination of Euro-American society at least since the defeat of General George Armstrong Custer's attacking 7th Cavalry at the Greasy Grass—better known as the Little Big Horn—in June 1876.

Initial encounters between Euro-Americans and the peoples of the Northern Plains tended to be peaceful—indeed, the American Fur Company's trading post in southeastern Wyoming (Fort Laramie) regularly traded with the Oglala for buffalo hides during the 1830s. But serious tensions began to develop in the 1840s as throngs of White settlers bound for Oregon became increasingly disruptive of Indigenous lifeways in the region. Of particular concern to the Lakota, Northern Cheyenne, and other itinerant[26] peoples was interference with the buffalo herds. Buffalo were plentiful enough on the Plains to provide in abundance for the needs of the various peoples who relied upon them for their survival, but their numbers belied the fragility of this relationship. As Richard White notes, buffalo migratory patterns were unpredictable so that, despite their great numbers, they were not always ready to hand (White 1978: 335). Moreover, once a hunt was underway the herd would stampede, the greater endurance of the buffalo allowing them eventually to outdistance the hunters' horses. And there was the danger that, even if it did not reduce the herds themselves, continuous hunting might have the effect of permanently driving them away to safer ranges. As Thomas Biolsi explains, this resulted in the development of a cooperative survival strategy wherein Lakota bands, though they spent the winter months in separate encampments, would come together in the summer to perform the Sun Dance and to take part in large communal buffalo hunts. This allowed both for a limiting of the number of hunts as well as the organization of a coordinated descent upon the herd by all hunters so that each would have a chance to make a kill before the inevitable stampede began (Biolsi 1984: 151).

All of this was put in jeopardy, however, by the ever-increasing presence of Euro-Americans on the Northern Plains. The flood of emigrants bound for Oregon, which intensified through the 1840s, swelled in 1848–49 as would-be prospectors flocked to the California gold rush. At the same time, large-scale hunting by Euro-Americans seeking hides began to alter the balance between the buffalo and the Indigenous peoples of the Northern Plains. Inasmuch as these encroachments jeopardized Lakota access to the buffalo, they were a threat to survival. Underscoring the seriousness that the Lakota attached to the imperative of restricting and regulating the number and conduct of buffalo hunts is the provisional policing function performed by the *akicita*.[27] As Biolsi observes, the institution of the *akicita*, whose instructions could not be disobeyed by even the most influential *itancan*, "stands in marked contrast to the generally noncoercive nature of Plains political relations" (Biolsi 1984: 151). The immoderate conduct of the Euro-American interlopers, then, violated even the most deeply held Lakota norms of reasonable social behavior—most particularly in their apparent disregard for the imperative of maintaining the balance of the sacred hoop.

In 1851, the United States concluded a series of treaties with the Lakota and other Plains peoples at Fort Laramie (which had become a U.S. Army post two years earlier), undertaking to safeguard them "against the commission of all depredations by the people of the . . . United States."[28] For their part, the Indigenous signatories pledged thenceforth "to make restitution or satisfaction for any wrongs committed . . . by any band or individual of their people, on the people of the United States, whilst lawfully residing in or passing through their respective territories."[29] But the peace was short-lived and tensions mounted once more as the Army proved either unwilling or unable to curb the disruptive excesses of Euro-Americans on the Plains. In particular, the buffalo herds were being decimated for their hides and, increasingly, by sport hunting. An organized resistance was finally aroused—principally following the lead of Red Cloud—in the mid-1860s when the Army began to construct a network of forts along the Bozeman Trail, which passed through important Lakota hunting grounds in present-day Wyoming. In what came to be known as "Red Cloud's War," the Oglala fought back, wiping out a troop of some 80 soldiers near Fort Phil Kearney in December 1866 as a prelude to repeated attacks on the new Army posts the following summer. In the face of this opposition, the United States relented, concluding another treaty at Fort Laramie in 1868 in which it was agreed that the Bozeman Trail would be abandoned and the forts closed.

The 1868 Fort Laramie Treaty designated all territory of present-day South Dakota west of the Missouri River as the "Great Sioux Reservation"— roughly half of the total area of the state including the *Paha Sapa* (Black Hills), held sacred by the Lakota. Moreover, the treaty explicitly provided that none but the Lakota themselves would enjoy any right of passage over or settlement upon these lands. In addition, substantial adjacent areas of Montana, Nebraska, North Dakota, and Wyoming were acknowledged as "Unceded Indian Territory." In return, the Lakota signatories pledged to make their permanent homes on the reservation, venturing off only to hunt buffalo in agreed areas and only for so long as the herds ranged there in numbers sufficient to justify such excursions. All of these promises came to ruin, however, when Custer led an expedition into the Black Hills and discovered gold there in 1874. When the Lakota rejected subsequent attempts by the United States to buy the *Paha Sapa*, U.S. President Ulysses S. Grant ordered the Army not to enforce the prohibition against Euro-American incursions into Lakota territory. Predictably, prospectors began registering claims in the Black Hills in what rapidly became a new gold rush. By 1876, the stage was set for a catastrophic confrontation after the U.S. Congress suspended the distribution of rations (guaranteed under the 1868 treaty) until such time as the Lakota agreed to the sale of the Black Hills. This policy left many with no choice but to leave the reservation and join bands living in the traditional manner, always careful to evade the reach of the troops dispatched to "bring them in."

The War Department of the United States responded by ordering all Lakota bands back to the reservation, casting as outlaws and "hostiles" all those who refused to obey. A three-pronged Army expedition was dispatched under the commands of Generals George Crook, John Gibbon, and Custer with orders to intercept the recalcitrant bands and deal with them by force. On 25 June, Custer—already notorious in Indian Country for the 1868 massacre of a sleeping Cheyenne village on the Washita River—famously met his end at the Greasy Grass after his 7th Cavalry attacked a sprawling encampment of Lakota, Cheyenne, and Arapaho who, along with notables like Crazy Horse, Gall, and Sitting Bull, had left their reservations. But this rousing victory over the forces of the United States was fleeting. In response, the military was placed in direct control of the Lakota reservation. Sitting Bull and Gall were forced to flee into temporary exile in Canada while Army units under the command of General Nelson Miles mounted a relentless winter campaign to drive the remaining Lakotas back to the reservation. The surrender of Crazy Horse in May 1877 and his murder at the hands of soldiers attempting to arrest him four months later punctuated the end of large-scale Lakota resistance.

The violences visited upon the Lakota by the U.S. Army were but one factor in a much more thoroughgoing assault on their traditional lifeways, carried out both by design and by happenstance. The destruction of the buffalo herds, for example, was driven principally by the profit motive. But as the dire consequences this had for the Plains peoples became readily apparent, they developed into an explicit end in themselves. Indeed, the Euro-American buffalo hunters seem often to have fancied their own decimation of the buffalo herds to be the definitive answer to "the Indian question."[30]

Equally devastating, however, was the discursive construction of savagery in consequence of which it was possible to contemplate an "Indian question" in the first place. This was always in evidence but, not coincidentally, was invoked with moral fervor whenever Indigenous people(s) moved to defend themselves and their lifeways, as Standing Bear would later recall: "When we tried to protect ourselves, they called us 'savages,' and other hard names that do not look good in print" (Standing Bear 1975: 74). The casual advocacy of genocide that so often accompanied this issued as readily from sites of conspicuous social privilege as through the popular culture media of dime novels and the like. No less a figure than Theodore Roosevelt, only five years before he was to become president of the United States, told a New York audience:

> I don't go so far as to think that the only good Indians are the dead Indians, but I believe nine out of every ten are, and I shouldn't like to inquire too closely into the case of the tenth. The most vicious cowboy has more moral principle than the average Indian. Turn three hundred low families of New York into New Jersey, support them for fifty years in vicious idleness, and you will have some idea of what the Indians are. Reckless, revengeful, fiendishly cruel, they rob and murder, not the cowboys, who can take care of themselves, but the defenseless, lone settlers on the plains.   (Quoted in Hagedorn 1930: 355)

And together with these pernicious constructions of the Lakota and other Indigenous peoples, enduring ideational structures and commitments—for instance, Christianity and Western notions of "progress"—worked to make the wars prosecuted against them and the deliberate destruction of their traditional lifeways seem entirely natural, necessary, and desirable.

Ironically, the very fact that it engendered a worldview that was not conducive to the formation of enduring social divisions of class and authority furnished something of the basis upon which the unity of the hoop was ultimately to be assailed. Absent social class and orders of authority, the Lakota were not recognizable by Euro-Americans as constituting anything akin to a bonafide political community. As Pierre Clastres points out, a people without conspicuous social division whose "chiefs" lacked the authority to rule could occupy only one category in Euro-American taxonomy: that of the savage (Clastres 1994: 120–21). Of course, the Lakota were by no means the first people of the Americas to be inscribed in like manner by Europeans. Time and again, Indigenous peoples were measured against European referents—broadly, "civilization" and Christianity—and judged inferior to the extent that they seemed poorer on these accounts. And this, of course, became the crux of much of the justificatory rhetoric attending the prosecution of wars against Indigenous peoples and the seizure of their lands. At the same time, every new victory in the colonial conquest of the Americas seemed to Europeans to confirm the veracity of their initial estimations of their Indigenous victims (Berkhofer 1978: 24). The ascription of savagery to all the peoples they encountered also contributed to the homogenization of the various peoples of the Americas into the singular category of "the Indian," notwithstanding that sustained contact ought to have created an awareness of significant cultural and linguistic differences between Indigenous peoples (Berkhofer 1978: 24).

Also interesting in this context, is the fact that virtually every instance of significant resistance by Indigenous peoples has been attributed almost exclusively to the cunning and machinations of some extraordinary leader—Tecumseh, Pontiac, Crazy Horse, and Geronimo, to name a few—as though without their allegedly uncommon abilities their respective peoples could not and would not have orchestrated any appreciable opposition. Some, like Sitting Bull, were cast in the role of the guileful agitator, seeking to exploit any opportunity to arouse rebellion among their respective peoples. In 1890, the Plains saw the arrival from Utah of the Ghost Dance, the central rite of a new hybrid faith expounded by a Paiute prophet named Wovoca. Combining aspects of Christianity with Indigenous beliefs, Wovoca foretold the imminent appearance of a Messiah who would restore the devastated buffalo herds and drive the Europeans from the continent. This fundamentally peaceful movement—dubbed the "Messiah Craze" by alarmed Euro-Americans—soon developed a considerable following among the Lakota. Although, he did not personally subscribe to the new faith, Sitting Bull was assumed by Euro-Americans to be behind what they perceived as mounting unrest on the Plains reservations. An editorial appearing in *Harper's Weekly* in October 1890 declared, "In the present state of affairs the noted Sioux chief Sitting Bull, who has already been the source of so much trouble in the course of Indian affairs, appears once more as a prominent figure."[31] Indian police in the employ of the federal Bureau of Indian Affairs were dispatched to arrest Sitting Bull at his home, whereupon a struggle took place with members of Sitting Bull's band leaving several dead, Sitting Bull among them.[32]

Apparently, the Euro-American assumption of the inferiority of "the Indian" had made it difficult to accept that what seemed to be a significant organized expression of resistance could have arisen absent the contrivance of some exceptional leader/agitator. Of course, there is nothing especially bewildering about this view given the broader conceptual terrain upon which it was constructed. After all, if Indians comprised an inferior race existing in the wilds of the continent without the attributes of anything Euro-Americans might recognize as the marks of political community, small wonder that nothing short of the idea of the exceptional individual could sufficiently explain those instances when their collective conduct seemed unified to a single purpose. Indeed, it would seem that this very view, in a most extreme form, underlay the call to genocide that appeared in an editorial in *The Aberdeen Saturday Pioneer* (a weekly newspaper in Aberdeen, South Dakota) following Sitting Bull's death:

> The proud spirit of the original owners of these vast prairies inherited through centuries of fierce and bloody wars for their possession, lingered last in the bosom of Sitting Bull. With his fall the nobility of the Redskin is extinguished, and what few are left are a pack of whining curs who lick the hand that smites them. The Whites, by law of conquest, by justice of civilization, are masters of the American continent, and the best safety of the frontier settlements will be secured by the total annihilation of the few remaining Indians.[33]

This bizarre oppositional rendering of the noble and ignoble savage in advocacy of genocide drew on a long tradition of essentializing discourses about Indigenous peoples. It also reflects the ambiguity of these discourses, which quite readily underwent reversals as befit the varied exigencies of colonial conquest and domination—here, Sitting Bull, only two months earlier described in *Harper's Weekly* as "the source of so much trouble in the course of Indian affairs," is remembered as noble and proud of spirit when it seems to suit a call to exterminate surviving Indigenous people. Nine days later the 7th Cavalry moved to oblige, opening up its Hotchkiss guns on the Mnikowoju encampment at Wounded Knee.

Even before the dawn of the reservation era, determined efforts were dedicated toward the cultural assimilation of the Lakota and other Indigenous peoples. These were eventually formalized and institutionalized through a range of measures, negotiated and/or imposed. The 1868 Fort Laramie Treaty, for example, included a number of material incentives for the abandonment of traditional lifeways, not least that an annual annuity to be paid in goods to all eligible persons under the treaty would be doubled for those who ceased to "roam and hunt" in favor of a sedentary agrarian existence.[34] For many Euro-Americans, the persistence of Indigenous languages stood as a serious impediment to assimilation. Indeed, in its official report, the Indian Peace Commission of 1868 lamented that the interests of peace might better have been served had more been done to press the use of the English language in place of the "barbarous dialects"[35] of the Plains peoples: "Through sameness of language is produced sameness of sentiment, and thought; customs and habits are moulded and assimilated in the same way, and thus in process of time the differences producing trouble would have been gradually obliterated."[36] At the same time, Christian missionaries endeavored to impart—and implant—Europe's monotheistic faith commitments in place of the Indigenous spiritual traditions that they regarded as paganistic.[37] Arbitrary prohibitions were also made against the performance of central rites of Lakota cultural and spiritual traditions, as exemplified in the federal ban on the Sun Dance imposed in 1904.

In the face of Euro-American westward expansion in the nineteenth century and the formidable ideational structures that sustained it, then, the Lakota people were confronted with two choices: genocide or ethnocide. They could accept assimilation or, for a time at least, they could break out from the shrinking reservations and join the defiant bands that, while always trying to evade the U.S. Army, adhered as best they could to traditional ways. The latter option, however, became ever less viable in light of the apparent willingness of the U.S. Army to employ violent means without restraint, as at places like Sand Creek,[38] Washita, and Wounded Knee. Mirroring the genocide/ethnocide gamut confronting Indigenous people(s) was a concomitant debate in the dominating society about how best to deal with the "Indian question." Few thought the preservation of traditional lifeways to be either a practical or desirable option; in the minds of most Euro-Americans, termination in one form or another was a foregone conclusion. Instead, the debate turned on the question of whether Indigenous people could be brought into the "civilized mainstream" of Euro-American society or were utterly incorrigible in this regard: the choice, then, boiled down to assimilation versus extermination. Those who preferred the latter found their interests expressed best by the Cavalry. By the late-nineteenth century, though, the hopes of the assimilationists lay in the growing number of Lakotas who, having been deprived of their traditional means of subsistence by the systematic extermination of the buffalo, settled on the reservations in close proximity to the federal Indian Agencies where they could receive the food rations guaranteed to them under treaty.

Of course, few imagined that assimilation could be completely effected in the immediate term, expecting instead that it would likely have to await the passing of at least the last generation to have lived in the traditional way. But even if the "savage" could not be exorcised, at a minimum the Lakota had to be incorporated. It was one thing for Euro-Americans to be able to extend their laws over the territory of which they had dispossessed the Lakota, but it was quite another to impose them on Lakota people themselves. So long as defiant bands resisted settlement on the reservation, the precise number and identities of the persons in those bands could not be known. The same was true of all others who resisted sedentarism and persisted in their itinerant tradition, even when they confined their movements to reservation lands— Euro-American authorities still wanted for fixed addresses where persons known to them might reliably be found. This confounded attempts by Euro-Americans to hold many Lakota traditionalists accountable to their laws, reinforcing the imperative of assimilation. Most fundamentally, it called for the destruction of the essential sociopolitical unit around which traditional values and lifeways were organized: the *tiyospaye*. Accordingly, the foundational paradigm of assimilation turned on the making of the modern individual.

If disciplining practices on the reservation lacked the panoptic gaze that Foucault would deem essential to their perfection, Biolsi argues that other aspects of Lakota social relations were equally problematic from the point of view of assimilationists. In particular, they defied the commodification of land and labor (Biolsi 1995: 29). This only exacerbated the irreducibility of the *tiyospaye* to the indispensable basis of Euro-American socioeconomic organization: the household founded upon the male-headed nuclear family. The challenge confronting the advocates of assimilation was, as Biolsi puts it, that "[t]he Lakota had to be forced to conform to a certain minimum definition of modern individuality" and thus "be constituted as social persons who could fit into the American nation-state and the market system of metropolitan capitalism" (Biolsi 1995: 30). In virtually every significant regard, these needs were inconsonant with a worldview derived from the idea of the sacred hoop as well as the distinct lifeways and forms of political economy it sustained.

A confluence of private empropertyment, assimilative residential schooling for Lakota children, and the institution of a "blood quantum" system for determining who was Lakota and who was not served the aims of the assimilationists and the Indian Agents alike while undermining the *tiyospaye*. If itinerant lifeways could be extinguished at the same time as individual interests were forged around immovable property, social control would be better enabled by the vulnerability of those interests.[39] At the same time, private empropertyment, especially of land, threatened (or promised) to unsettle the unity of the *tiyospaye* by engendering and institutionalizing particular material interests at the level of the individual. The 1887 General Allotment Act— also known as the Dawes Severalty Act in connection with its principal sponsor, Massachusetts Senator Henry L. Dawes—sought precisely these ends. According to Stephen Cornell, "[b]y distributing tribal lands to tribal members, assigning each allottee a tract of homestead size, and granting

US citizenship to allottees, the Dawes Act set out to destroy tribe as territorial, economic, and political entity" (Cornell 1988: 33). Moreover, as Biolsi points out, "private property under capitalism—instituted in the Native American case by the allotment of Indian lands in severalty—presupposes the state as a protector of the common interests of individual property owners" (Biolsi 1995: 31). Thus, the making of the propertied individual was simultaneously a means by which to broaden the legitimacy of state authority while eroding traditional social relations and structures of governance.

For its part, blood quantum has arguably been the most disruptive influence on the *tiyospaye*, its divisive legacies enduring at the root of much present-day factionalism on Lakota reservations. Based on the degree of Lakota blood as determined by lineage, this quantitative measure was institutionalized as the means by which to determine who was genuinely Lakota and therefore entitled to treaty benefits. In 1917, blood quantum became the basis for differential treatment in respect of allotted lands and their proceeds—those with "less than one-half Indian blood" were judged competent to manage their own affairs of property while all others remained official wards of the government whose interests were held in trust (Biolsi 1995: 41). Over time, the ascriptions "full-blood" and "mixed-blood" were internalized by the Lakota themselves as important identity markers. As Biolsi argues, this should hardly be surprising given their ubiquity in the everyday as definitive arbiters of the material exigencies of life (Biolsi 1995: 41).[40] A perverse effect of this, of course, was that it not only undermined the *tiyospaye* but incited mutually agonistic social relations as well. These tensions found an institutionalized site of contention after a liberal-democratic representative form of local "self-government" was imposed under the U.S. Indian Reorganization Act of 1934 (IRA). Antithetical to Lakota traditions of sociopolitical organization characteristic of the *tiyospaye*, these "tribal councils," which have become the officially sanctioned (by the U.S. Department of the Interior through its Bureau of Indian Affairs) political-administrative authority on reservations, are regarded by many as "puppet governments" of the advanced colonial settler state.[41]

What were perhaps the most pernicious ethnocidal practices were reserved for Lakota children: the violences of residential schools, suffered by young people from Indigenous communities throughout the United States and Canada even beyond the mid-twentieth century. The housing of students at boarding schools served the immediate function of moderating the behavior of their parents on the reservations whence they could offer them no protection (Hannah 1993: 427–28). Their manifest purpose, however, was to assimilate and, toward that end, every effort was made to sever students' connections to the traditions of their parents.[42] And as Birgil Kills Straight has recalled of his own school days, the standard school curriculum was but one of the violences visited upon students:

> Most of us have experiences in Bureau of Indian Affairs schools where we were whipped, flogged, and were taught to learn that the father of this country was George Washington. We have seen in the textbooks ourselves as Indians being savages or heathens out raiding nice innocent people moving across our land, raping.   (Kills Straight 1977: 177)

Not surprisingly, the hegemonologue spoke these (re)presentations in the language of the colonizers; foreshadowed in the 1868 report of the Indian Peace Commission, bans on students' use of their own languages were the norm in the schools. Amply illustrating the seriousness of these prohibitions, Standing Bear tells of having sought special permission to speak to his father (who understood no English) when he came to visit him at the Carlisle Indian School in Pennsylvania (Standing Bear 1975: 149). But in spite of such harshly conceived and enforced strictures, the assimilationists remained certain of the rightness of their convictions—Dawes himself paid approving tribute to the Carlisle school as "the earliest and most persistent in developing the industrial faculties of Indian pupils" (Dawes 1899: 284).

(Post)colonial Legacies

M. Annette Jaimes notes that education has been the preferred approach of those who, thinking it only proper that the virtues of civilization be shared, fancy their ethnocidal inclinations to be benevolent: "The task confronting those who would better [Indigenous peoples'] miserable lot" from this perspective "is thus fundamentally educational, to acquaint them with all they are 'missing' through their obstinate insistence on remaining 'outside of history' " (Jaimes 1992: 8). But such pretensions could issue only from the prior rendering of Indigenous people(s) as primitive, a view very much consistent with the commitments of a host of Western academic orthodoxies, if most conspicuously that of Anthropology. As Jaimes argues, the orthodox treatment of the aboriginal condition of Indigenous North Americans as a "Stone Age" existence—that, in the Europe of 15,000 to 40,000 years ago, is reckoned to have been stark indeed—misleadingly constructs them against a European referent:

> Since the implements and utensils employed by American Indians at the time of first contact with Europeans were made mainly of stone, Eurocentric

orthodoxy—both popular and scholarly—has always decreed that their situation in life *must* have equaled that of Europe during its Stone Age. To be blunt, the assumption is that not only were the Indigenous peoples of America retarded by at least ten millennia behind the levels of material and other sorts of cultural attainment already reached in Europe, but they were also physically and intellectually incapable of favorably altering this situation without the intervention of Europeans. (Jaimes 1992: 9; emphasis in original)

Of course, there is something of a *coup de force* in this inasmuch as the considerable achievements of Indigenous societies[43] are unequivocally denied in a move enabling the oppositional construction of a civilized European/ Euro-American Self. Small wonder, then, that such (mis)representations should be as central to contemporary expressions of Americanness[44] as they were in the travelogues of the first Europeans in the Americas.[45] And when articulated through the material violences of colonialism/advanced colonialism (e.g., as in the schooling of the "savage") their central function and most perverse effects are simultaneously realized. We should thus never lose sight of the fact that discourses of Americanness are discourses of domination.

It is, of course, essential that such processes and practices of domination be recounted—they are, after all, inseparable from accounts and experiences of the colonial encounter in the Americas. But unintended consequences may also follow from a recitation of the violences of colonialism/advanced colonialism, even in so cursory a fashion as I have done here. Specifically, this runs the risk of once again constructing Indigenous peoples monolithically, defining them this time by their victimization. As Robert Craig reminds us in this context, "[d]omination, cultural or otherwise, is rarely total, in spite of the intentions of the powerful" (Craig 1997: 4). Again, the colonized are not reducible to "passive skin" awaiting the inscriptions of the colonizer. And to render them as though they were/are, even unintentionally, is to repeat the founding assumption of the idea that the aboriginal condition of the Indigenous people(s) of the Americas was a Stone Age existence wanting for the arrival of Europeans and, with them, civilization. We are bid to remember, then, that the colonized have agency and this begets strategies of resistance and survival that, in turn, enable both survivances and creative remakings of individual and collective Selves. In short, though domination might be catalytic, it is not by itself definitive of the subjectivities of colonialism's Others.

As Bhabha would have it, hybridity, forged in the play of assimilation and resistance, is an enduring legacy of the colonial encounter. According to Rice, "Lakota adaptability was expressed in various apparent compromises, especially religious ones" (Rice 1991: 30). Similarly, William K. Powers argues that shifting and intermixing their traditional spiritual commitments with Christianity was a common strategy for Lakota people negotiating the violent contradictions of life in the contact zone, so that aspects of each could variously be summoned in ways that "satisfied the needs of the particular time and its attendant crises" (Powers 1990: 140). A nonlinear, nonhierarchical view of the world was particularly well suited to such strategies since it did not necessitate that different belief systems be mutually exclusive. As Scott J. Howard explains, this is how it is that there is no necessary contradiction inherent in Black Elk's adherence to both Catholicism and Lakota traditionalism:

> Black Elk's network of possibilities, because it did not rest on actual propositions or on doctrinal statements of belief, was inclusive—capable of including many rituals and practices—whereas the Christian network is basically exclusive, regarding other religions and systems of thought as incorrect or "pagan."   (Howard 1999: 119)

But we should also take care not to miss the important role of colonial violences here. Rice notes that, "when Neihardt asked Black Elk why he became a Catholic in 1904, he answered simply, 'my children had to live in this world' " (Rice 1991: 30). The Catholic overtones in Black Elk's accounts of Lakota spiritual traditions thus bear witness to the "conditions of occupation" under which, according to Rice, "he attempted to make survival conditions for Lakota consciousness and self esteem favorable in an overwhelmingly Christian world" (Rice 1991: 12).

Of course, hybridity does not develop in uniform fashion. That is to say, some among the colonized will, more readily than others, adopt and adapt the lifeways and perspectives of the colonizers. Said notes that "[l]arge groups of people believe that the bitterness and humiliations of the experience which virtually enslaved them nevertheless delivered benefits—liberal ideas, national self-consciousness, and technological goods—that over time seem to have made imperialism much less unpleasant" (Said 1993: 18). Others in their own communities may be less persuaded, or to varying degrees. This reflects the circumstance that assimilative practices are typically uneven in application, inconsistent in implementation, and fraught with internal contradictions.[46] Moreover, the prismic workings of intersubjectivity, mediating the transmission of the hegemonic narratives and ideas of the colonizers, confound the fixing of stable meaning. This too results in ambivalent outcomes that often elude the colonizers' intentions and designs. It also reveals an important source of divergent responses to colonialism among the colonized themselves.

The heterogeneity of survival and resistance strategies often profoundly marks the political landscapes of postcolonial communities. In the cases of Indigenous North American peoples it tends to manifest most conspicuously in the cleavage between so-called progressives and traditionalists. This division, which has long been the crux of political discord in many Indigenous North American communities, reflects two broad approaches to resistance: one focused principally upon survivances of pre-colonial lifeways; the other employing what Bhabha might describe as a strategy of mimicry, deriving from decidedly local readings of the hegemonic texts of the colonizers. Both approaches necessarily work through and evince the hybridity that is an inescapable feature of the postcolonial condition. What might in the first instance seem to be a sharp oppositional dichotomy between those who have been assimilated and those who have not, thus turns out to be a site of mutual ambivalence at the liminal interstices between the idealized Self and Other identities of the colonial encounter.

In the political life of Lakota communities, tensions between progressives and traditionalists have long figured prominently. The latter adhere, so far as is possible, to traditional Lakota values, spiritual beliefs, and lifeways. Many traditionalists have withheld recognition of the authority of the federally recognized IRA tribal councils, most fundamentally on grounds that they are a Euro-American form imposed from without and, being hierarchical structures of decision-making authority organized around executive offices, are antithetical to traditional Lakota sociopolitical structures. The legitimacy of the councils has not, of course, been at all helped by recurring concerns about corruption[47] or by a fairly widespread sense among traditionalists that they serve as conduits through which advanced colonial domination and exploitation are actualized and sustained. Consequently, for the traditionalists, authority is deferred to traditional elders, though they do not enjoy any official sanction from the constituted authorities of the dominating society.

Contemporary contestation around the matter of political allegiances, then, reflects long-standing disputes that arise most fundamentally from the heterogeneity of survival and resistance strategies. In the earliest days of the reservation era, Lakotas who settled close to the Indian Agencies where they could receive their treaty rations were sometimes disparagingly called the "Hang-Around-the-Fort-Indians" by members of the defiant bands that more actively resisted incorporation and assimilation; more recently, though with equal flavor of reproach, some progressives have been branded "Uncle Tom-ahawks" for their willingness to support the tribal councils. For their part, traditionalists are regarded by some as hopelessly archaic and out of step with contemporary realities—a disposition that, in the view of its critics, amounts to a futile renunciation of "progress" as well as any claim to a share in the material rewards that embracing the values and commitments of the dominating society seems to promise.

From time to time, these disagreements have manifested in fearsome bouts of political violence. In the early 1970s, for example, considerable bloodshed followed an upsurge of traditionalist political activism led by members of the American Indian Movement (AIM) on the Oglalas' Pine Ridge Reservation. There was nothing particularly novel about the issues involved: principally, corruption in Tribal Council and the Bureau of Indian Affairs and the failure of the United States to honor the 1868 Fort Laramie Treaty. But AIM invested these and other matters with a renewed moral charge through appeals to traditional values and lifeways, and this did not sit well with supporters of the Tribal Council and others who neither identified with traditionalism nor shared AIM's sense of historical grievances.[48] Against this backdrop, the period 1973–76 saw a sustained wave of political violence, targeted primarily against traditionalists. Decades later, controversy still surrounds allegations that dozens of AIM activists and their supporters were murdered, principally by the loosely organized Guardians of the Oglala Nation (GOONs) loyal to then-IRA Tribal Chairman Richard Wilson.[49] As Pine Ridge resident Gladys Bissonette described the situation at the time:

> We have been harassed on our Reservation, Pine Ridge. The United States Government is using Chairman Richard Wilson and his regime of goons to intimidate the Indians. We cannot touch a gun or even keep a .22 in our homes to go hunting. No, they take them away, but the goon squad has highpowered automatic rifles to shoot at people. (Bissonette 1977: 176)

Though so intense a campaign of terror has not been seen since, incidences of politically motivated violence remain a fact of life for many on Pine Ridge. Indeed, fears that the violence of the 1970s might be re-ignited followed the January 16, 2000 takeover of the reservation's Tribal Offices by a group calling itself the Grass Roots *Oyate* (People). The group's principal objectives were to expose what they charged was ongoing corruption by the IRA government and to replace it with a traditional elder-centered form based on consensus decision-making.[50] In the first few months of the occupation of the Red Cloud Building, a spate of bomb threats was received, a council member was beaten in Rapid City, threats against specific people were broadcast on the reservation's radio station, and the home of prominent Oglala journalist Charmaine White Face was ransacked in what appeared to be a deliberate attempt at intimidation.[51]

Incidents such as these reflect the colonial/advanced colonial disruption of traditional values and lifeways in two important and interrelated respects. Most tangibly, they are telling as to the extent of fragmentation of the *tiyospaye* effected by measures such as the IRA and the blood quantum regime. In the making of the modern individual, traditional modes of sociopolitical organization and their attendant mechanisms for dispute resolution were severely undermined. Thus, Cornell links the reservation era loss of the ability to secede to the contemporary factionalism in many Indigenous North American communities (Cornell 1988: 38). At the same time, the imposition of centralized hierarchical structures of political authority under the IRA promoted division as influence fell disproportionately to different individuals and groups. Importantly, all of this took place in the context of the harsh conditions of reservation life—traditional bases of material reproduction having been destroyed, competition for increasingly centralized power was made all the more acute as it became a *sine qua non* for survival.[52] Factionalism manifesting in the opposition of traditionalists and progressives has been further complicated by crosscutting fragmentation pitting "full bloods" against "mixed bloods." And all of this, of course, is quite revealing of the second and more fundamental sense in which the colonial experience has worked its violences to considerable effect: the cosmological commitment to the transcendent unity inherent in the idea of the sacred hoop has apparently lost at least some of its salience for many.

Enduring Traditionalism

In spite of all of this, however, traditionalists continue to constitute a significant proportion of the Lakota *Oyate*. Of course, Spivak's concerns about the idea of the traditional as the basis for a new locally hegemonic subjectivity imposed over and against its own marginalized Others are worth remembering here. Likewise, we might be given pause to consider the social constructivist view that the traditional is always in some measure an invention inspired by the exigencies of the present. But if, as I have argued, all history is current history, and if we accept the notion that hybridity is an inescapable part of the postcolonial condition, the politics surrounding this view itself are at least as significant as those it purports to expose. That is to say, the observation that all communities are imagined communities[53] gives comfort to conservative political agendas if allowed to flatten alterity and liberate the privileged from the still-unfolding histories of domination and exploitation upon which their privilege is founded. Though all history might be current history, it is not without its enabling antecedents, and its violences, like its rewards, are not apportioned evenly. The caution I would offer here, then, is the same one implied in Spivak's insistence that strategic essentialism, while indispensable to an oppositional politics, must be invoked only in a "scrupulously visible political interest" (Spivak 1988b: 205). In this sense, it is vital that we bear in mind that, while there is a distinct epistemic and political community identified with Lakota traditionalism, not all Lakota people would count themselves as members of this community.

What, then, can we say of traditionalists? Most fundamentally, in the contestations over Lakota identity they settle on a form that fits with the cosmological commitments described earlier. And though this necessarily be speaks a particular politics, it is political in the same sense as all ideas associated with the pursuit/preservation of the good life, but not necessarily as a highly mobilized set of ideas oriented toward the realization of some particular social objective(s)—put another way, traditionalism itself is not necessarily programmatic, though some traditional people are moved to political activism. But even among the politically active, few would stake an idealized claim to some "pure" tradition. As we have seen, hybridity inevitably results from survival strategies as colonized people devise and improvise their own identities and commitments in contesting and adapting to those forced upon them.[54] But years before postcolonial scholars began to discuss this idea, Lakota traditionalists were already quite clear on the matter. A glossary of terms appended to a June 1974 declaration of Indigenous sovereignty, for example, included this text under the heading "Traditional":

> [Traditionalism] does not mean returning to the lifestyle of four-hundred years ago—an obvious impossibility—nor does it mean denying the real benefits of Western technology. . . . Traditionalism implies for us a unity and inter-relatedness of all the areas of human endeavor, i.e., religion, science, medicine, politics, etc. Further, it means that we hold the earth to be sacred, not to be bought and sold, and that use of certain areas of land belongs to those people who are part of that land. All other forms of life that are part of an area of land have equal rights with humanity.[55]

Much more than some contemporary conjuring in the ethnographic present, then, traditionalism is perhaps best understood as being organized around important survivances of values and lifeways from what might be called an aboriginal condition, but not as a wholly unaltered inheritance.

While traditionalism need not manifest in political activism, enduring fidelity to the values, lifeways, and cosmological commitments outlined in this chapter is nevertheless discernible in many of the urgent causes to which traditional people have lent their support. The efforts of the Grass Roots *Oyate* to replace the IRA government of Pine Ridge with a traditional form give us one such glimpse. So does Charmaine White Face's compelling rejection of the blood quantum regime as antithetical to traditional ties of kinship.[56] Equally telling is the ongoing dispute over the *Paha Sapa*. In 1980, the U.S. Supreme Court ruled that lands guaranteed to the Lakota in the 1868 Fort Laramie Treaty—including the Black Hills—had been taken illegally in an 1877 Act. The Court upheld a 1979 ruling ordering compensation in the amount of $17.5 million plus annual interest at a rate of 5 percent from 1877, for a total exceeding $100 million.[57] Remarkably, in spite of the intense poverty on their reservations, Lakota people refused the settlement, which was subsequently placed in trust and continues to accrue interest.[58] Most Lakota people regard the settlement monies as nothing more than an attempt by the United States to legitimize the theft of the Black Hills, which they insist are not for sale—a cynical exercise whose antecedents include unsuccessful attempts to cajole individual Lakotas into signing treaties relinquishing title on behalf of the whole of their people.[59]

There is perhaps no more dramatic expression of enduring traditional Lakota cosmological commitments than the efforts mounted on behalf of the buffalo since the late 1990s. Led principally by artist and activist Rosalie Little Thunder, a small group of Lakota traditionalists has been active in a campaign to save Yellowstone buffalo from the Montana Department of Livestock. Though protected in the park, the buffalo sometimes venture out seeking new foraging areas during the winter months and this has been a matter of some concern for Montana cattle ranchers who fear they might be financially harmed if the buffalo transmit disease to their herds. In response, the Department of Livestock has tried hazing and, when that fails, has resorted to shooting the wayward animals. Consistent with the ideas expressed as *mitakuye oyasin*, Lakota traditionalists regard the buffalo as relatives, as the *Tatanka Oyate* (Buffalo Nation or Buffalo People). Though a buffalo economy based on communal hunts was central to their aboriginal lifeways, these were highly ritualized practices founded on a commitment to the idea that the buffalo willingly surrendered their bodies to the needs of the hunters who, completing the covenant, undertook not to kill beyond the satisfaction of their needs—again, a relationship founded upon the imperative of balance. But in the winter of 1996–97, the Yellowstone herd was roughly halved by a combination of especially difficult weather and, for the greater part, through slaughter by government agencies (LaDuke 1999: 152). Ever since, traditional people and their supporters have worked to raise awareness, suffering the elements and even arrest in their attempts to intervene and stop the killing.

Traditional cosmological commitments to the idea of the sacred hoop and its transcendent unity and interdependence loom large in all of this. As Lakota spiritual leader Arvol Looking Horse cautions, "if there is no buffalo, then life as we know it will cease to exist" (quoted in LaDuke 2000: 66).

As we have seen, and as the title of this chapter suggests, the colonial encounter has witnessed continuity and change in Lakota lifeways. But there has also been something of both continuity and change in colonialism itself. In terms of the former, Lakota traditionalism can no more be abided today than it could more than a century ago. But though colonialism endures, it has been perfected into an advanced form. Likewise, the particular forms of its violences have undergone a fundamental change in emphasis. Where once the coercive apparatus of the settler state was brought persistently to bear in the suppression of Indigenous tradition, today that is the exceptional case. Instead, advanced colonialism is characterized most immediately by the violent erasures performed by the hegemonologue. Insisting upon the implausibility of Indigenous peoples' accounts of their traditions, the hegemonologue underwrites the unyielding common senses of the dominating society. What this represents, of course, is more a transformed emphasis than a *sui generis* process of domination—the hegemonologue, in one form or another, has spoken itself over and against Indigenous peoples' attempts to name themselves and their destinies since the earliest days of the colonial encounter. The hegemonic knowledges it bears continue to inform the impressions of Indigenous peoples held by Euro-Americans as well as the latters' attitudes toward contemporary struggles around land claims and other issues. Representation has thus become foremost among the violent practices sustaining advanced colonialism. And as we shall see in chapter 5, it functions no less effectively in this regard for want of instrumental design.

Notes

1. Seven related groups make up the Lakota: Hunkpapa, Itazipco, Mnikowoju, Oglala, Oohenunpa, Sicangu, and Sihasapa.
2. Walker was the physician at Pine Ridge Agency on the Oglala Lakota reservation from 1896 to 1914.
3. Tyon was also a principal informant himself (Powers 1986: 125).
4. See, for example, Rice (1991).
5. The conventions of Western academic writing become quite problematic here. There is no accepted means of referencing texts that would give direct authorial credit to Tyon and Black Elk and, moreover, it might be regarded as a culturally insensitive mistake to try to individually ascribe oral literatures in this way. Their voices are nevertheless the sources of the texts published by Walker and DeMallie.

6. Underscoring this point, most readers likely will not be able to tell simply from reading their names in references that many of the contemporary scholars cited in this book are Indigenous people themselves.

7. This, of course, can never be completely divorced from the unequal power relations of colonialism.

8. The rendering of "*itancan*" as "chief," for example, misses a great deal of meaning and, as we shall see, bears inappropriate and misleading implications as well.

9. See Parks and DeMallie (1992).

10. Indeed, it is precisely this mistake that Bedford and Workman (1997) argue is at the root of Crawford's (1994) reading of the Great Law of Peace of the Haudenosaunee Confederacy as a security regime.

11. Genesis 1:28 (Revised Standard Version) reads, "And God blessed them, and God said to them, 'Be fruitful and multiply, and fill the earth and subdue it; and have dominion over the fish of the sea and over the birds of the air and over every living thing that moves upon the earth.' "

12. Here, then, is the site of struggle at which people have sought, through both practical and spiritual means, to adjust to the transcendent balance of the natural order of which they are part.

13. See, for e.g., Severt Young Bear's testimony before U.S. Federal Judge Warren Urbom at the "Sioux Treaty Hearing" at Lincoln, Nebraska in December 1974, a transcript of which can be found in Dunbar Ortiz (1977: 124–34).

14. Again, absent linearity, the identification of difference is not easily translated into hierarchy.

15. Although certainly not underwritten by the same stratified social relations upon which the Athenian *polis* was constructed.

16. According to Standing Bear, "the council made no laws that were enforceable upon individuals. Were it decided to move camp, the decision was compulsory upon no one. A family, or two or three of them, might elect to remain in the old village. However, in most matters a decision that was favorable for one was favorable for all" (Standing Bear 1978: 129–30).

17. See Price (1994).

18. Catherine Price (1994: 451) also notes that "Euroamerican emissaries were overwhelmingly unsuccessful in inducing [*itancan*] to render decisions without the general consent of their council peers."

19. Lewis Bad Wound explained it thus to the "Sioux Treaty Hearing" in 1974: "Our system of government is the exact opposite from what the United States is. I would describe theirs as a pyramid with their heads of state at the top. Ours is the exact opposite. These people tell us what to do. We don't tell them what to do" (Bad Wound 1977: 184).

20. Any who might be tempted to read the revenge complex as a mark of primitivism in itself might be reminded that avenging the November 1940 German bombing of Coventry is frequently counted among the motives for the February 1945 bombing of Dresden—a city swelled with refugees and with virtually no military significance—by British and U.S. bomber forces. Through much of the 1990s,

U.S. air strikes against Iraq were commonplace in retaliation for a variety of unwelcome behaviors by that country. Similar examples abound. It is also noteworthy that the logics of strategic deterrence turn on the explicit threat of revenge.

21. For a plausible hypothesis as to the original sources of conflict, see Biolsi (1984).

22. Large-scale exterminative warfare seems not to have been a feature of the aboriginal condition of most Indigenous peoples of the Americas. Indeed, this inspired some scorn from European soldiers in instances when they allied with them. One British officer complained thus of his Narraganset allies: "This fight is more for pastime, than to conquer and subdue enemies. . . . They might fight seven years and not kill seven men" (quoted in Utley and Washburn 1977: 43).

23. See chapter 5.

24. As Joan Laxson points out, "For many, the movies, TV, and comic strips have firmly established the stereotype of the stoic Plains Indian with a warbonnet. There seems to be little that can compete with this image, and it begins quite early in life" (Laxson 1991: 376). The result is that the cultural symbols of the Plains peoples have become pan-Indian markers of "authenticity" in the dominating society: "Postcards and souvenirs depicting Indians almost always show them costumed in beaded buckskin and feathered warbonnets, even in parts of the country such as New England, where this was not typical native dress. . . . These pan-Plains symbols are semiotic markers that reassure the tourist audience that, indeed, this is an 'authentic' Indian experience" (Laxson 1991: 378).

25. On the Wounded Knee massacre, see Utley (1963); challenging Utley's account of the sources of the massacre, see Ostler (1996).

26. As Roxanne Dunbar Ortiz points out, the common ascription of nomadism to the Lakota is erroneous and misleading: "The people had summer and winter homes, and moved in relation to the herds in a definite pattern and in a very organized way" (Dunbar Ortiz 1977: 68).

27. As Beatrice Medicine has explained it, "if in a buffalo hunt, one hunter went out before the entire group went, this jeopardized the welfare of the whole people" (Medicine 1977: 121).

28. "Treaty of Fort Laramie with Sioux, Etc., 1851," in Kappler (1971: 594).

29. "Treaty of Fort Laramie with Sioux, Etc., 1851," in Kappler (1971: 594).

30. See, e.g., Cook (1967) and Mayer and Roth (1958).

31. "Three Noted Chiefs of the Sioux," *Harper's Weekly* (October 20, 1890), p. 995.

32. The Ghost Dance was extinguished on the Plains with the massacre of Spotted Elk's band at Wounded Knee two weeks later.

33. *Aberdeen Saturday Pioneer*, December 20, 1890, cited in Venables (1990: 37). It is worth noting that the author of this casual call for genocide who was the editor of the *Aberdeen Saturday Pioneer*, was none other than L. Frank Baum who would later write *The Wizard of Oz*.

34. Article X reads, in part: ". . . the sum of ten dollars for each person entitled to the beneficial effects of this treaty shall be annually appropriated for a period of

thirty years, while such persons roam and hunt, and subject to the revision of the Secretary of the Interior, shall be binding on the parties of this treaty, twenty dollars for each person who engages in farming to be used by the Secretary of the Interior in the purchase of such articles as from time to time the condition and necessities of the Indians may indicate to be proper" ("Fort Laramie Treaty of 1868," reprinted in Dunbar Ortiz 1977: 96–97).

35. Cited in the 1887 report of J.D.C. Atkins, Commissioner of Indian Affairs, excerpted in Crawford (1992: 48).

36. Cited in the 1887 report of J.D.C. Atkins, Commissioner of Indian Affairs, excerpted in Crawford (1992: 48).

37. See Tinker (1993).

38. On November 29, 1864, several months after Black Kettle was instrumental in securing peace from the Cheyenne War, his band's encampment at Sand Creek was attacked by an Army command under Colonel John M. Chivington. Even after Black Kettle hoisted a U.S. flag together with a white flag of truce over his tipi, Chivington's soldiers did not relent, retiring only after killing and mutilating the bodies of all those who did not make good their escape.

39. See Hannah (1993).

40. See also Strong and Van Winkle (1996).

41. For an engaging scholarly exchange on this point and on the pros and cons of the Indian Reorganization Act more generally, see Washburn (1984,1985), Biolsi (1985).

42. While serving as school superintendent for the State of New York in the late-nineteenth century, for example, zealous assimilationist Andrew S. Draper actively promoted boarding schools over reservation day schools with the explicit aim of disrupting relationships between students and their families (Johnson 1974: 79).

43. See Weatherford (1988).

44. See chapter 5.

45. See chapter 6.

46. The implementation of policies designed to sedentarize Indigenous people in North America, e.g., typically proceeded through the allotment of land that was entirely inhospitable to agriculture.

47. For a former Tribal Treasurer's first-hand account of the corruption and political opportunism—implicating both the Tribal and U.S. Government—she encountered and worked to expose, see White Face (1998).

48. See Roos et al. (1980).

49. The exact number of deaths remains unsettled, but has variously been put at between 58 and 69. See Churchill (1994: 197–205) and Johansen and Maestas (1979: 83–84). The U.S. Federal Bureau of Investigation, which has jurisdiction over capital crimes on reservations, disputes both the numbers claimed as murders and allegations that it has failed to seriously investigate many of the deaths. In a recent report, the Justice Department lists the names of 57 persons killed.

Ambiguous circumstances surround a number of the deaths not ruled homicides. Of those that the report attributes to foul play, several have never resulted in a conviction. According to the report, two of the dead were killed by Deputy U.S. Marshals in separate incidents at a roadblock near Wounded Knee in April 1973; two other deaths are attributed to shootings by BIA police officers. See United States, Department of Justice, "Accounting for Native American Deaths, Pine Ridge Indian Reservation, South Dakota: Report of the Federal Bureau of Investigation, Minneapolis Division" (May 2000).

50. Personal communication, February 20, 2000.
51. White Face believed the vandalism was intended to "send [her] a message" and that it represented an attack on traditional people more generally since traditional items in the home seemed to have been specifically targeted for destruction. Charmaine White Face, personal communication, October 28, 2001. See also Melmer (2000).
52. Shannon County, which accounts for most of Pine Ridge Reservation, is frequently counted as the poorest in the United States.
53. See Anderson (1983).
54. Joane Nagel and Matthew Snipp describe the voluntary dimension of this change as "ethnic reorganization," arguing that "an ethnic group's response generally involves some attempt to choose among alternatives, some degree of decision-making, some effort to reorganize for survival" (Nagel and Snipp 1993: 206).
55. "Declaration of Continuing Independence by the First International Indian Treaty Council at Standing Rock Indian Country, June 1974," in Dunbar Ortiz (1977: 203).
56. See White Face (2001).
57. For an account of the proceedings authored by the son of the attorney who litigated the case, see Lazarus (1991).
58. More than two decades later, the total held in trust has grown to well in excess of a half billion dollars. Frederic J. Frommer, "Black Hills Are Beyond Price to Sioux," Los Angeles Times (August 19, 2001).
59. See Thin Elk (1977: 157).

PART 2

(Re)Presentation

CHAPTER 5

Advanced Colonialism and Pop-Culture Treatments of Indigenous North Americans

I f the preceding chapter is not easily reconciled with some readers' expectations about the Lakota past and present, that is likely because it is quite strikingly at odds with the stock representations of Indigenous people(s) more generally that have long been a staple of Euro-American popular culture. The manner in which most members of the dominating society understand Indigenous North Americans is in the aggregate, as "Indians." This marker is somewhat unique in the dominating society, being at once reverent and pejorative—an odd sort of conceptual schizophrenia that takes on a unitary form in the idea of the "noble savage." Although most are well aware that cultural and linguistic variations differentiate Indigenous peoples, there is also a persistent core of beliefs about presumed basic commonalities that serves to reinforce the aggregate construct; indeed, the all too common zero-plural rendering of "*the* Indian" as a collective identity marker for Indigenous North Americans bears clear testament to this effect. This well-established and deeply embedded way of perceiving Indigenous North Americans in the dominating society is also accompanied by both spatial and temporal boundaries in the popular imaginary about "the Indian." The fixing of such parameters is the inevitable extension of practices of objectification and, to the extent that it confines Indigenous North Americans to remote spatial and temporal contexts, it is central to sustained practices of advanced colonial domination. These discursive confinements, which are borne in both literary and nonliterary textual forms of representation as well as through varied technologies of iconography and performance, are consumed in the dominating society and Indigenous communities alike—though they certainly are not received as unproblematically in the latter as in the former, important elements are internalized in both contexts nonetheless.

The Figural-Discursive Construction of Indigenous North Americans

If discourse is the amalgam of statements that, taken together in texts, become imbued with meaning beyond what they connote in the singular, figurally discursive re-presentation pertains to the subtexts of meaning borne by communicative forms that rely less on language and more on depiction through the deployment of forms and semiotic markers.[1] Like linguistic discourses, figural forms function via the construction and selective arrangement of markers that, in turn, draw upon a vast store of commonly held, though usually unenunciated, ideas, beliefs, and understandings. In Foucault's formulation, discourses are to be regarded not "as groups of signs (signifying elements referring to contents or representations) but as practices that systematically form the objects of which they speak" (Foucault 1972: 49). Likewise, figural-discursive representation involves the reciprocal-simultaneous[2] construction and presentation of objects, with little or no acknowledgment of the inherent subjectivity of that process. Importantly, the subtext of ideas advanced in presentation concerns not only how the object is to be perceived, but also makes implicit claims regarding the appropriate relationship between subject and object. As John G. Galaty puts it, "the narratives condensed in popular images serve to situate their subjects politically and to define how their destinies should or must be fulfilled," such that "moral stances towards 'imagined' subjects and justifiable strategies of action towards 'real' subjects are defined" (Galaty 2002: 351). It is also worth emphasizing that none of this need be undertaken instrumentally in order to have the same effect. In fact, as Loomba points out, the fictive characters of literature may variously, even simultaneously, (re)produce narratives of both domination and resistance that bear no necessary connection to authorial intent (Loomba 1998: 74).

Ample illustration of this point is found in the 1995 film, *The Indian in the Cupboard*, jointly released by Columbia Pictures and Paramount Pictures. Based on the Lynne Reid Banks novel of the same name, the film tells the story of a 9-year-old boy named "Omri" (Hal Scardino) who discovers that the combination of an old cupboard and an apparently enchanted key has the power to bring his toys to life in miniature. But there is a catch: the transformed toys are animated by real persons, magically snatched across time and space from their own lives to become captive surrogates for Omri's otherwise lifeless figurines. So it is that "Little Bear" (Litefoot), a fictional member of the very real and venerated Onondaga Wolf Clan, is plucked from his life in the year 1761 when Omri unwittingly subjects a plastic representation of an Iroquois man to the magic of the cupboard. Following on an initial—if fleeting—trepidation on the part of Little Bear, he and Omri develop mutual trust and esteem as the two embark upon an adventure that has Omri learn lessons about friendship, loyalty, and respect for others.

In general, *The Indian in the Cupboard* does quite well in living up to the promises of its promotional press as a "touching tale" and "terrific family entertainment." Moreover, the film's makers seem to have made a genuine effort to step outside some of the more pervasive stereotypes about Indigenous North Americans—appropriate to an Onondaga, for example, Little Bear rejects Omri's offer of a tipi in favor of the materials with which to construct a longhouse. In another scene, an effort is also made to problematize the sanitized depictions of wholesale slaughter of anonymous, one-dimensional Indigenous North American characters that were the stock and trade of an earlier genre of cinematic screenplay, the so-called western. But despite these laudable moves, other moments see the Little Bear character reduced to spectacle in noteworthy ways.

The rendering of spectacle is a very common phenomenon in Euro-American pop-culture treatments of its Others, and one that is by no means confined to the cinemagraphic medium. Michael Dorris tells of having come upon an unlikely display in a souvenir shop in the Cook Islands: "perched in a prominent position on a shelf behind the cash register was an army of stuffed monkeys, each wearing a turkey-feather imitation of a Sioux war bonnet and clasping in right paw a plywood tomahawk" (Dorris 1987: 98). Dorris's encounter with this spectacle on a South Pacific island may seem improbable until one takes account of the remarkable capacity of Western—and, in particular, American—popular culture to serve as a vehicle for a plethora of constructed images of Others, variously defined in racial, ethnic, political, socioeconomic, and gender terms. Though undoubtedly more benignly conceived and certainly less offensive in outward appearance, the "Indian" in Omri's cupboard is not entirely different from the vulgar caricatures on the shelf in the souvenir shop. The Iroquois figurine in the movie started out as a lifeless object, an exemplar of the pretension of the dominating society to appropriate, (re)interpret, and exploit Indigenous North American cultures and their various markers. Like the monkeys in the imitation war bonnets, its primary instrumental function was to provide amusement, thereby rendering as spectacle the culture and society upon which it drew while simultaneously objectifying the real people who are part of that culture and society. Even after the transformation in the cupboard, Little Bear remains something of a curiosity and, not insignificantly, is wholly dependent upon and subject to the whims—however well intentioned—of a child.

It is in the fundamental presentation of the appropriate relationship between subject and object—that is, the treatment of Little Bear as both spectacle and chattel—that *The Indian in the Cupboard* makes its inadvertent contribution to the objectification of Indigenous North Americans, denying them independent ontological significance. As if to underscore this point, the packaging that accompanied the film's eventual release by Columbia TriStar Home Video bore the promotional announcement that an "Indian," cupboard, and key were included therein free of charge.[3] In partial fulfillment of this promise, a small plastic figurine, similar to the one transformed by Omri's magic key and cupboard, accompanied videocassettes offered for sale. The "Indian" in the cupboard thus became the property and plaything of real children. As such, each figurine is discursive to the extent that its very function implies a specific relationship between subject and object—in this case: owner and property; spectator and spectacle.

Though it surely is not the product of any deliberate machination, this aspect of the "Indian" figurines nevertheless shares effect with more instrumentally conceived technologies. Among the best known of these was Buffalo Bill's Wild West. Bill Cody's traveling exhibition of live performances, which was described in an English newspaper of the day as "reality" (Simon and Spence 1995: 69), toured the United States and Europe at the end of the nineteenth century, mythologizing moments in the westward expansion of Euro-American society. Presenting dramatized commemoration as historically authentic representation, Cody cloaked his subjective interpretations and portrayals of real events and real people(s) in claims of objective truth. In so doing, he and his show invigorated and reinforced their audiences' stereotypically derived beliefs and assumptions about the comportment and lifeways of Indigenous people(s), to say nothing of the "national" mythologies of the audiences themselves. And by presenting Indigenous people(s) as an undifferentiated mass, always, in the end, going down to inevitable defeat in mock battle against the cavalry, the figural/performative narratives of the Wild West show objectified and eulogized them in the same instant, thereby conveying a sense of inevitability, even progress, about their forced subjugation.

A similar—if rather less ostentatious—construction is noted by Christina Klein: namely, photographs of the aftermath of the infamous massacre of unarmed Lakotas by elements of the U.S. Army's 7th Cavalry at Wounded Knee, South Dakota on December 29, 1890. In particular, those taken by photographer George Trager were commercially distributed throughout the United States and saw sufficiently wide circulation as to be readily subsumed under the rubric of popular culture. Characteristically, these images starkly juxtaposed the orderliness of the uniformed troops detailed to clear the site of the massacre with the haphazardly strewn corpses of the Lakotas, many of which had been frozen into decidedly disorderly and grotesque poses. As Klein argues, it is important to note that this contrast was a product of deliberate composition:

> [T]hese photographs articulated the cultural ideology of their historical moment and acted as visual expressions of a concern for order, control, and hierarchy. Yet . . . their power derived in part from their ability to render this ideology almost invisible. In their realism, their seeming fidelity to nature, the photographs appeared to be simply found rather than constructed, to be objective truth rather than point of view.   (Klein 1994: 56)

These photographs, then, were an exercise in figurally discursive (re)presentation. As Klein puts it, "Trager's photographs articulated the contemporary logic of creating social order by consigning distinct types of people to their proper spaces" (Klein 1994: 56).

Much less subtle was the so-called Sioux War Panorama—a series of paintings, billed as an accurate representation of events surrounding the abortive Santee uprising in the summer of 1862 that culminated in the mass public hanging of 38 Santee men. The Panorama, a kind of crude mechanical forerunner of the motion picture in which a collection of scenes ran in sequence on a large canvas scroll, toured the American Midwest with its owner/creator, John Stevens, during the 1860s and 1870s. However, neither the visual representations nor the accompanying narration established the context of the uprising—that is, of Santee people who, having been dispossessed of most of their land, faced starvation as well as indifference to their plight on the part of settlers and government officials alike. Instead, as John Bell describes it:

> The performance began: the crankist advanced the canvas roll, image after image, while Stevens recited his "correct" version of events which had occurred only recently and not very far away; his show redefined the 1862 Sioux uprising for the settler audience as an epic narrative of white innocence, Indian savagery, vulnerable nature, and death. Stevens's audience was already familiar with the uprising, and probably already believed in the moral ideology with which Stevens's panorama defined and framed the events. But the occasion of watching Stevens's performance in the company

of other settlers allowed the audience, as a whole, to define what had happened up in the Lake Shetek region as another chapter in a vast American mythic history, a history whose ideological function was to justify white acts of retribution against "Indian savagery."   (Bell 1996: 279–81)

Here linguistic and figural discourse worked in tandem, vigorously developing the objectified image of the Santee as savages until catharsis was delivered by a depiction of the mass hanging, followed by an idyllic scene in which settlers enjoyed the pleasures of the countryside, apparently unburdened of any further fear of danger.

Each of these nineteenth-century forms, Cody's Wild West, Trager's Wounded Knee photographs, and Stevens's Panorama, ultimately performed the same political function. All three exploited the representational technologies of their time as means by which to advance nonlinguistic (or not exclusively linguistic) narratives about Indigenous people(s), constructing them as Other in myriad ways. At the same time, each made implicit claims about Euro-American society—a constructed and preferred image of some Other, after all, only makes sense in conjunction with some image of the Self, which is similarly constructed and preferred. In the broader context of Euro-American self knowledges, Jimmie Durham argues that "America's narrative about itself centers upon, has its operational center in, a hidden text concerning its relationship with American Indians" (Durham 1992: 425). In each of the nineteenth-century spectacles described above, that hidden text is laid bare. As Dorris argues, the emphasis on violent conflict with Euro-Americans that commonly predominates in depictions of Indigenous North Americans is "clearly at odds with well-documented facts" and "only serves to reinforce the myth of Indian aggressiveness and bellicosity and further suggests that they got what they deserved," while simultaneously conferring honor upon the Euro-Americans who defeated such a dangerous foe (Dorris 1987: 100–01).

Western philosophy's linear conceptions of being are here betrayed by the constructions of multiple and intersecting binary oppositions. The dichotomous renderings of Self and Other are mapped over by a host of others, including but not limited to inside/outside, masculine/feminine, culture/nature, literacy/orality, and civilized/savage. The former pole in each of these oppositions bespeaks a knowing subject, the latter its objects. Accordingly, each is inextricably bound up with relations of power so that domination/subordination may also be read from them. In their reduction of alterity to simple dualisms, they deny ambiguity and hybridity, mystifying the complexities and contingencies of identity, all the while underwriting the organization of essentialized categories of people into pre-formed hierarchies. Moreover, as V. Spike Peterson argues,

> not only do oppositional constructions distort the contextual complexity of social reality, they set limits on the questions we ask and the alternatives we consider. True to their "origin" (Athenian objectivist metaphysics), the dichotomies most naturalized in Western world views (abstract-concrete, reason-emotion, mind-body, culture-nature, public-private) are both medium and outcome of objectification practices. Retaining them keeps us locked in to their objectifying—reifying—lens on our world(s) and who we are. (Peterson 1992: 54)

That these dichotomies have enabled often conflicting Western accounts of the colonial encounter underscores their indeterminacy. Loomba, for example, notes that portrayals of the colonies as naked women dominated by their European masters readily gave way to a reversal of the rape narrative, casting women colonists as vulnerable to the sexual predations of Native men (Loomba 1998: 79). And underscoring both their contingency and their grounding in the material conditions of colonial domination, this shift typically accompanied moments of heightened resistance or rebellion by the colonized. Paranoid fantasies about the abduction of Euro-American women by Indigenous raiders of the vulnerable frontier homestead—an almost requisite part of any pop-culture account of the nineteenth-century "Wild West"—thus correspond with the final, most mythologized struggles to extend Euro-American suzerainty across the whole of the North American continent. The apparently carefree women pictured in the concluding frames of Stevens's War Panorama thus signaled the categorical defeat of the Santee uprising—a triumph of culture over nature, reason over appetite, and civilization over savagery.

The centrality of such renderings to the (re)production of colonial domination cannot be overstated. They are also inseparable from the construction and maintenance of the settler state—a form whose uniqueness has not been well appreciated by students of International Relations. In Benedict Anderson's celebrated formulation, the nation is an "imagined community," a collective idea, instantiated through the everyday performance of cultural scripts, from which the state derives its legitimacy (Anderson 1983). But, having no long-standing attachment to the territories on which they reside, settler states derive their national mythologies more from their own creation struggles than from the claim to some ancient common heritage. Typically, this takes the form of epic tales of heroic labors in service of an inspired new vision of the just society. In the case of the United States, one body of such tales involves the Revolutionary War for independence from Britain. All but eclipsing this, however, are the legion accounts of the struggles to civilize the untamed wilderness of the North American continent. Indeed, even in the Declaration of Independence no less a figure than Thomas Jefferson penned the following words: "HE [King George III] has excited domestic Insurrections amongst us, and has endeavoured to bring on the Inhabitants of our Frontiers, the merciless Indian Savages, whose known Rule of Warfare, is an undistinguished Destruction, of all Ages, Sexes and Conditions."[4] Thus, Cody's traveling show, to take a single example, was but one moment in what has become a bona fide tradition of celebrating the mythologies of the "Wild West." Here, the perennial preoccupation with the dangers posed by "hostile Indians"—hostile by nature much less than by reason—has played well to the Lockean-inspired sense that the right of property was conferred in the making of improvements that would turn the land to productive purposes (Locke 1924: 132): not only had Indigenous peoples let much of the continent lie fallow, but they also posed a threat to those who would now bring forth the plough. By making it safe for settlement, cultivation, and commerce, therefore, those who answered the call to "go West" could find the legitimacy of their claim to the lands they entered in the very act of subduing the Indigenous peoples they encountered there. It was thus in the "taming" of the West that that West was "won."

Following Anderson and noting Charles Tilley's observation that the state rarely achieves congruity with the nation, Campbell argues that, because states "do not possess prediscursive, stable identities" and yet must reconcile their claims to legitimacy with a host of identity formations, they "are never finished entities" (Campbell 1992: 12). For Campbell, the discursive construction of danger, in opposition to which the state's identity is represented, is an essential element in its (re)production. Thus, "[t]he constant articulation of danger through foreign policy is . . . not a threat to a state's identity or existence; it is its condition of possibility" (Campbell 1992: 13).[5] So too, the rendering of "the Indian" as dangerous can be seen as integral to the (re)production of the settler state: indeed, it is central to the national mythologies of settler states in general, and to those of the U.S. in particular. In this sense, sports team mascots derived from Indigenous imagery and identity markers, Indian head nickels, and the like might also be read as the functional equivalents of war trophies, demonstrating simultaneously the danger faced and the superiority of the forces that overcame it. This places them in the same tradition as Cody's Wild West, Stevens's Panorama, and Trager's photographs. By objectifying Indigenous North Americans and presenting them in the aggregate as spectacle, always in the context of abortive and catastrophic conflict with Euro-American society, each of the nineteenth-century spectacles also served to conflate Indigenous people(s) with the natural challenges offered up by the "untamed" terrain on the American frontier. Thus, even if some among the audiences might have found it lamentable, the colonial subjugation of Indigenous people(s) was made to seem every bit the inevitable—even desirable—outcome of "progress" and the steady march of "civilization."

But lest we succumb to hubris and begin to think ourselves very much more sophisticated than our forebears, consigning the misuse and abuse of images of Indigenous people(s) to the "bad old days," it is perhaps sobering to take note of others who have betrayed such pretensions in the past. Writing in the *Atlantic Monthly* more than a century ago, George Bird Grinnell observed that "[m]ost thoughtful people believe that in the past the Indians have been greatly wronged by the whites, but imagine that this is no longer the case" (Grinnell 1899: 260). Yet, some ninety-five years later, Steve Hayward could still comment that "[n]ot too far gone are the days when any country museum served the naïvely colonial role of representing Indigenous American cultures to all who ventured in" (Hayward 1994: 109). Writing in the mid-1990s, Hayward himself might well have been offered a caution against the dangers of presuming one's own times to be advanced beyond the misguided or distasteful ideas and conventions of some earlier era. Of particular concern to him are the dioramas that figured so centrally in the museum exhibits of (presumably) bygone days. But Hayward would no doubt be displeased to learn that even now the diorama continues to be employed as a means by which to present Indigenous people(s) to members of the dominating society.

On a visit to Toronto's Royal Ontario Museum near the time of this writing, I encountered a striking example of the objectification and compartmental-ization of Indigenous North Americans via a combination of linguistic and figural discourse wherein the diorama featured prominently. The discursive tax-onomy of the museum's various galleries was immediately apparent. Accorded status as bona fide civilizations, galleries celebrated the cultural achievements of, among others, Ancient Egypt, Nubia, the Greeks and Etruscans, Imperial Rome, and Byzantium. Europe, of course, was represented in a series of several dedicated galleries. Tucked away one floor below street level was a small gallery identified as "Indigenous Peoples Exhibitions," which turned out to house a temporary exhibit of the work of three contemporary Anishnabek artists. To my surprise, the museum's permanent gallery devoted to the expo-sition of Indigenous people(s) was a small room next door identified as "Ontario Archaeology." The discursive impact of this is inescapable: Indigenous people(s) were not presented as constituting civilizations in their own right but, rather, as the subject matter of a Western academic discipline. Even more astonishing, however, was the sign identifying that subject matter that, placed at the entrance to the gallery, read "Ontario Prehistory." Nowhere had Mesopotamia or Ancient Egypt been identified as "prehistoric." Yet, even with exhibits purported to be characteristic of the mid-seventeenth century, the whole of this gallery was subsumed under a classificatory term reserved in the popular imaginary to dinosaurs or the ancestors of homo sapiens. Clearly, then, it would be difficult to argue that temporal markers account for this discrepancy in the application of a discourse that elicits a sense of nascent or primeval form. Rather, the usage here is rooted in the foundational myths upon which the Western academic disciplines were created—in particular the objectivist strategies that authorized the division of the peoples of the world into those who have history by virtue of their having formed vibrant and ever-evolving civilizations, and those whose presumed undynamic social systems made them the concern of Anthropology.

If the linguistic discourse identifying the "Ontario Archaeology Gallery" was disquieting, the figurally discursive representations inside the gallery seemed even more so. Apart from a few small scale models of ancient fortresses and a mock-up of an Ancient Egyptian grave site, the diorama was not employed as a means by which to elaborate the history, culture, or lifeways of the civilizations exhibited on the upper floors of the museum. And while I frequently found the collection and display of grave relics, ancient art, and other artifacts—in short, the cultural heritages of societies long ago conquered and plundered of these treasures—somewhat unsettling, they generally were presented, in and of themselves, as the momentous achievements of dynamic societies. By way of contrast, the centerpiece of the "Ontario Prehistory" exhibit was a series of four full-scale dioramas, complete with life-size mannequins purportedly depicting moments in the daily lives of Ontario's Indigenous people(s) at different intervals throughout history. Complementing the "prehistory" discourse, the first diorama depicted a mammoth kill in Beringia—a considerable distance from Ontario—12,000 years ago.[6]

The diorama, Umberto Eco tells us, "aims to establish itself as a substitute for reality, as something even more real" (Eco 1986: 8). Hayward is highly critical of the diorama as a representational medium, noting that its characteristics "are telling in regard to the dialectics of representation and colonization facilitated by commodification" (Hayward 1994: 109). As he puts it:

> The diorama elicits the charm of an artifact masquerading as something built by invisible, supremely objective, quasi-divine hands. It suggests an

unmediated and pristine correspondence between its subjects and what, within the terms of the epistemology operant in the production of the diorama, objectively exists in a real, outside world. . . . The lack of any obvious nod to the subjectivity and interestedness of the maker is indeed suspicious.   (Hayward 1994: 109)

If the diorama in general implies an objectivity that is not truly extant, the dioramas in the "Ontario Prehistory" exhibit seemed all the more problematic. Consistent with their placement in the "Ontario Archaeology Gallery," each was presented in two parts: in the background stood the mannequins, frozen in ostensibly typical depictions of some earlier time; in the foreground was a cut-away of a mock excavation, complete with stakes, notebooks, various tools, and partially unearthed artifacts. As discourse, the archaeological dig in progress in the front half of each diorama emphatically punctuated and completed what was initiated by the name of the gallery: the objectification of Indigenous North Americans as academic curiosity. But, more than this, it seemed also to affix the stamp of science, as if to verify the authenticity of the scenes depicted in the background. Thus, not only was the subjectivity of the diorama not acknowledged, but an implicit claim to objectivity was also made.

Popular Culture and Colonialism

Said defines imperialism as both the practice and ideology of colonialism (Said 1993: 9). Imperialism, in this view, extended beyond the actual colonization of the Americas to include the manufacture and diffusion of particular knowledges that allowed British troops garrisoned at Jamestown, for example, to rationalize their own presence there. Likewise, the ideological component offered up an accounting to British subjects in London for the maintenance of far-flung colonial possessions even when the costs of the endeavor sometimes outstripped the immediately tangible benefits. Thus, Said contends that a focus on material exploitation alone is not sufficient to develop an understanding of imperialism and colonialism. Rather, he proposes that "[b]oth are supported and perhaps even impelled by impressive ideological formations that include notions that certain territories and people *require* and beseech domination, as well as forms of knowledge affiliated with domination . . ." (Said 1993: 9; emphasis in original). Chief among these forms of knowledge are notions of inferiority and superiority.

Through his textual analysis of a number of the acclaimed works of Western literature, Said offers compelling insights into the extent to which the cultural paragons of the dominating society have, for centuries, been congenitally entangled with the business of colonialism. Sometimes with apparently benevolent intent, sometimes less so, much of Western literature has, at least implicitly, accepted and advanced the colonial project by virtue of its acceptance of the status quo of unequal power relations as the standard frame of reference. Consequently, it has rested on a foundation of constructed images of Others. The discourses that flow from these essentialized images are familiar—in their extreme form, they lead to claims that, for example, some given Others do not value human life the way We do. Writing on the adjustment of thought to purpose, E.H. Carr observed that "to depict one's enemies or one's prospective victims as inferior beings in the sight of God has been a familiar technique at any rate since the days of the Old Testament" (Carr 1964: 71). And this fulfills a function of legitimation, allowing Us to rationalize whatever We might undertake to do to Them. Said pays particular attention to the essentialized image of Arabs, long predominant in the West and emphasized in times of crisis, such as during the 1991 Gulf War: ". . . Arabs only understand force; brutality and violence are part of Arab civilization; Islam is an intolerant, segregationist, 'medieval,' fanatic, cruel, anti-woman religion" (Said 1993: 295). The outcome of this process of dehumanization is that it is made morally acceptable—if not imperative—to wage war against this Other, contradictions notwithstanding. This is a point that cannot be overemphasized, particularly given the criticism that too exclusive a focus on literary and other forms of representation can have the effect of obscuring the direct forms of violence unleashed in the colonial encounter.[7] What must be preserved, then, is a sense of the unbreakable connection between the violences of representation and those of the sword or musket—or the cruise missile, for that matter. As Loomba puts it, "[f]rom the very beginning, the use of arms was closely connected to the use of images: English violence in colonial Virginia, for example, was justified by representing the Native Americans as a violent and rebellious people" (Loomba 1998: 99).

It is important to recognize that there is much more of significance to this than a straight functionalist account like Carr's might suggest. Dehumanizing practices are simultaneously generative of both their subjects and objects. Recognizing this, Said treats the (Oriental) Other as productive of the (Western) Self:

The Orient is not only adjacent to Europe; it is also the place of Europe's greatest and richest and oldest colonies, the source of its civilizations and languages, its cultural contestant, and one of its deepest and most recurring images of the Other. In addition, the Orient has helped to define Europe (or the West) as its contrasting image, idea, personality, experience. (Said 1979: 1–2)

In this he acknowledges a debt to Foucault, treating Orientalism as a discourse through which the Orient and the Occident have been constructed in opposition to one another and such that knowledges about the Other are also necessarily knowledges about the Self. So, while J. Ann Tickner is quite right in noting, for example, that "[p]ortraying the adversary as less than human has all too often been a technique of the nation-state to command loyalty and increase its legitimacy in the eyes of its citizens . . ." (Tickner 1988: 436), it is essential that we recall as well the productive emphases found in Campbell's and Dorris's observations about articulations of danger. Said is also sensitive to both the productive and legitimizing dimensions of Orientalism, characterizing the discourse as "a created body of theory and practice in which, for many generations, there has been a considerable material investment" (Said 1979: 6).[8]

Curiously, however, Said scarcely makes mention of the Indigenous people(s) of the Americas. The cause of this omission may be that American global imperialism has served to obscure the fact of ongoing colonialism within the geographical confines of the United States itself. More likely, though, it is a function of Said's belief that "direct colonialism has largely ended" (Said 1993: 9). While this may be the case for most of Africa, Asia, and the Middle East, it is not so throughout the rest of the world—most certainly not in the Americas. To contend otherwise is to accept the status quo of the extant "international" system of states—itself a product of European imperialism—and thereby to reinforce the relative voicelessness and invisibility of peoples denied full and equal standing therein. It is, in short, to validate the ongoing direct colonial subjugation of the Indigenous peoples of the Americas. Still, Said's analysis of the ways in which the culture of an expansionist society is constitutive of colonial practices is instructive, particularly with respect to those places where direct colonialism has been sustained. The development and dissemination of the particular knowledges generated by this process have in no way been lacking with respect to Indigenous North Americans. Nor, as we have seen, have they been restricted to linguistic expression or the textual medium of written literature.

In the advanced colonial societies of North America, the incorporation into popular culture of images of Indigenous people(s) fulfills a further function not provided for in Said's analysis. To be sure, the legitimizing application is still operative, continually reproducing the requisite sense of inevitability or, at least, of necessity about the original conquest—after all, the idea that one's own society has been both perpetrator and beneficiary of aggression against other peoples is an unpleasantry that is all too easily suppressed in constructions of the past. In this sense, the cultural practices discussed by Said are not so much a matter of directly sustaining the colonial project as they are about reassuring members of the dominating society about their own past. This is not to say that exonerating the colonial heritage is not essential to the imperialist project, but of more immediate utility in advanced colonial societies is the maintenance of a sense among their members that the Indigenous inhabitants of the colonized territories they now occupy are no longer in existence or, at a minimum, that they are no longer subjugated under colonial domination. Thus, the "celebrations" of "vanished peoples" or "lost ways" through pop-culture semiotics—rendered as sports team mascots, for example—are one with the more explicitly malevolent representational forms discussed earlier. Discursively removed from mainstream spatio-temporal contexts, "the Indian" is exoticized, commodified, and consumed as unproblematically as any other mythical/mythologized people. With specific reference to the appropriation of Indigenous peoples' cultural and spiritual practices, Laurie Anne Whitt has observed that, "[w]hatever its form, cultural imperialism often plays a diversionary role that is politically advantageous, for it serves to extend—while effectively diverting attention from—the continued oppression of indigenous peoples" (Whitt 1995a: 5). Moreover, she submits that the very fact that so many of those who engage in the trade in Indigenous North American cultures—whether as producers, sellers, or consumers—"fail to recognize their behavior as reprehensible suggests that the diversionary function of cultural imperialism is operative at the individual level as well, where it deflects critical self-reflection" (Whitt 1995a: 6). This accords well with the notion that, in the main, the relationship between these practices and their effects is not an instrumental one—particularly at the level of the consumer.[9] However, this in no way diminishes the fundamentally oppressive nature of the resultant objectification of Indigenous people(s) so long as it contributes to the maintenance and reproduction of structures and practices of advanced colonial domination.

It is perhaps in the temporal and spatial confinement of Indigenous North America that popular culture has had its most profound effect. It is primarily by virtue of this function that the "diversionary role" to which Whitt refers is fulfilled. Ward Churchill notes that the largely constructed image of the archetypal Plains Indian, which has been delivered ad nauseam to mass audiences in movies and on television as a generic representation of Indigenous North Americans, has tended to be temporally bounded as well. Inasmuch as the bulk of such portrayals have been confined to an historical period spanning little more than a half century, he contends that "the Indian has been restricted in the public mind, not only in terms of the people portrayed (the Plains Nations), but in terms of the time of their collective existence (roughly 1825–1880)" (Churchill 1992: 232). The insistence on portraying Indigenous people(s) in this manner not only situates them squarely in the ethnographic present but furthers their collective objectification as curiosity, deflecting attention away from their contemporary struggles by making it seem as though all matters relating to their various relationships with the dominating society were, for better or worse, resolved long ago.

Each of the nineteenth-century spectacle forms discussed earlier served precisely this function. Even as the so-called Frontier era of American expansion drew to a close, Cody's Wild West, Stevens's Sioux War Panorama, and Trager's Wounded Knee photography, each in its own way, emphasized that the final settling of terms with Indigenous people(s) had, once and for all, been forged according to a provision with which their audiences were entirely familiar: the right of conquest as conferred by Native bellicosity. Each portrayed as matters of objective historical fact the decisive defeat and vanquishment of Indigenous people(s), while setting these images against unambiguous depictions of the ultimate triumph of the dominating society—the outcome of an apparently zero-sum game. In the case of Trager's photographs, Klein proposes that Victorian sensibilities would have been well served by the orderly assignment of space as, in one image, the burial detail lined up in tidy formation alongside a mass grave that contained the relative chaos of frozen Lakota corpses (Klein 1994: 56). As with the inevitable victory of the cavalry in Cody's Wild West and the final cathartic frames of the Sioux War Panorama, the discourse inherent in such images also seemed to preclude the possibility of Indigenous people(s) having an autonomous present or future.

In our own time, the dioramas in the Royal Ontario Museum's "Ontario Archaeology Gallery" function in similar fashion. The mock archaeological dig that physically separates museum visitors from the mannequins also relegates Indigenous people(s) safely to the (distant) past. This is reinforced by the discourse of "primitiveness" advanced via the representations made in the back half of each of the four dioramas.[10] At the same time, the display fixes both immediate and more general spatial parameters on Indigenous people(s). Like the corpses in Trager's photographs, the mannequins are visually framed within the confines of the exhibit itself. But, more significantly, they are assigned place—in this instance, what most members of the dominating society would be given to recognize as some nondescript tract of as yet "untamed" nature. Thus, Indigenous people(s) are once again conflated with the very landscape of days past. Seen in this light, the dioramas are indeed useful and instructive, teaching us more about the culture and society of which they are a part than that which they purport to depict.

Despite their best efforts to break with aggregate "Indian" stereotypes, the makers of *The Indian in the Cupboard* also do not forestall the predisposition to imagine Indigenous people(s) only in the past. Accordingly, the Little Bear character is delivered to Omri's cupboard from the year 1761. This does nothing to elucidate the contemporary struggles of real Onondaga people. Of course, it might well be advanced that an Onondaga of the eighteenth century also confronted the adverse effects of European colonialism and, to be fair, passing mention is made by the Little Bear character of the challenges associated with being interposed between the English and French colonies of that era. Nevertheless, though it may well have been benignly conceived, the choice of depicting an Onondaga of more than two centuries ago cannot be read as value-neutral in its ultimate effect—that is, in contributing to the popularly held notion that the colonial era exists only in the past. As Churchill argues:

> North American indigenous peoples have been reduced in terms of cultural identity within the popular consciousness—through a combination of movie treatments, television programming and distortive literature—to a point where the general public perceives them as extinct for all practical intents and purposes. Given that they no longer exist, that which *was* theirs—whether land and the resources on and beneath it, or their heritage—can *now* be said, without pangs of guilt, to belong to those who displaced and ultimately supplanted them. Such is one function of cinematic stereotyping within North America's advanced colonial system.   (Churchill 1992: 239)

Clearly, then, intent is much less relevant a consideration than effect in assessing whether a given representation of Indigenous people(s) is rightly regarded as a buttress to their continued colonial subjugation. And it is the dearth of portrayals of Indigenous North Americans in anything other than the distant past that has set the popular context such that each subsequent offering only serves to confirm what members of the dominating society already "know" about them. As Deloria has observed: "The tragedy of America's Indians—that is, the Indians that America loves and loves to read about—is that they no longer exist, except in the pages of books" (Deloria 1973: 49).

A Pretension to Pretend: The Appropriation of Voice

If instrumental design is not essential to the effectively oppressive nature of images of Indigenous people(s) in popular culture, it follows that even well-meaning people may unwittingly contribute to the maintenance of colonial oppression through their work. This is especially so in the case of non-Indigenous people who presume to speak in the voice of an Indigenous person, all protestations of honorable intent notwithstanding. Churchill locates the source of such literary misadventure in the legacy of colonialism itself:

> [T]he leading practitioners of identity appropriation are often motivated by sincere empathy and a genuine will to open viable channels of communication between oppressor and oppressed. The very real power dynamics of colonialism, however, preclude even the best and most sensitive of such efforts from doing other than reinforcing and enhancing the structure of domination at play. (Churchill 1992: 143)

The unequal power relations to which Churchill refers facilitate the remaking of the cultures, identities, and lifeways of Indigenous people(s) in such a way as to expedite their assimilation and consumption within the dominating society. Rose argues that the hegemony of distinctly Western epistemological assumptions of the universality of knowledge requires that non-European ideas be made to fit into places marked out in advance by Euro-derived intellectual traditions—that is, the ways in which they must be interpreted are, in many fundamentals, predetermined by particular ways of knowing (Rose 1992: 407).

Obviously, this has important implications with regard to members of the dominating society who would appropriate the voice of an Indigenous person. Leslie Marmon Silko argues that Anglo-American poets who believe that they have both the ability and the right to speak through their art as though they were Indigenous people necessarily proceed from two inherently racist assumptions: first, that a member of the dominating society is capable of unerringly thinking, feeling, and experiencing the world from the perspective of an Indigenous person; and, second, that such things as prayers, chants, and stories are not the cultural property of particular Indigenous peoples but, on the contrary, are in the "public domain" and may therefore be freely appropriated (Silko 1979: 211–12).[11] And, as Rose points out, the seemingly natural tendency of the works of such poets to conform to the stereotypically derived expectations of the dominating society has frequently enabled them to supplant real Indigenous people as the authoritative voices on their own cultures (Rose 1992: 412–14). What is at stake for Indigenous people, then, is nothing less than the preservation of their right to secure their own definitions of who they are.

But some, like J.C. Davies, take exception with the apparent ubiquity of critiques like Silko's. While Davies seems to concede that the blatant objectification of Indigenous people(s) in accordance with well-established stereotypes such as that of the "noble savage" is problematic, he raises the objection that Silko's denial of the possibility that a writer can scrupulously assume the voice of an Indigenous person "repudiates much of western literature as practiced within its own ethno-cultural milieu, and assumes that fiction in particular should be read as literal reportage on states of mind" (Davies 1994: 242). In this connection, Davies singles out two novels by Craig Lesley—*Winterkill* (1984) and *River Song* (1989)—in an effort to demonstrate that Silko's line of criticism is misdirected. In both novels, Lesley's protagonist is a contemporary Nez Perce man who, having been raised in the dominating society, encounters, explores, and embraces the heritage and traditions of his family and people. In attempting to give an authentic voice to this character within the tradition of literary realism and its putative objectivity, Davies suggests that Lesley may be vulnerable to criticism as "patronizing, unconscious of cultural relativity, and therefore implicitly racist" (Davies 1994: 233). Therefore, it is against the a priori assumption of objectivity bound up in literary realism and not the author's "racial-cultural incapacity to be realistic" that Davies thinks Silko's criticism should rightly be directed (Davies 1994: 243). Davies acknowledges that "the European novel has its genesis in part in a notion of 'human nature' that Silko is not alone in rejecting, and that Lesley, by employing pre-modernist techniques of realism such as plausible factuality and omniscient narrative, implicitly supports the notion of 'human nature' in his own work" (Davies 1994: 243). Nevertheless, he defends Lesley's pretension to give voice to his Nez Perce character:

> Lesley's characters exist as part of an historical process. They are particular, not generalized. Further, character in text, as point of view *and* acting, suffering narrative center, is a mode of depiction of ethics and values, and the fundamental convention of most novels—from *Pamela* to *The Color Purple*—is that one can "get inside" someone else; that is, a novel's explicit rhetoric stimulates such an operation and its readers expect it. To deny universal human nature is in any case as much an ideological stance as to affirm it . . . It would be sad if, as Euro-American understanding of the world were moving towards pluralism, some Native criticism were to take an ethnocentric and exclusivist position.   (Davies 1994: 243; emphasis in original)

In the end, then, it seems that Davies has, after the manner of Shakespeare's Marc Antony, come forward to praise Lesley's novels, not to bury them.

It may well be that Euro-American cultural *dabbling* is becoming more pluralistic, but it seems quite a considerable stretch to say that Euro-American *understanding* of the world is becoming more pluralistic, inasmuch as whatever new knowledges may be appropriated are subjected to the deforming constraints and impositions of decidedly Western epistemic commitments. If we extend this point into a consideration of differences in worldview in the context of differing ways of knowing, it would seem all the more poignant—especially given Davies's own earlier suggestion that Lesley's narrative technique "fosters, by its very clarity and attention to detail, the illusion of documentary objectivity" (Davies 1994: 233). Moreover, Davies's contention that Lesley's characters are particular and not generalized seems a weak and hollow point. For this to be of consequence, readers would require a more than passing familiarity with Nez Perce culture, values, lifeways, and so forth, in order to recognize characters as particular and not necessarily representative of a generalized aggregate of the people to which the author has ascribed them. Moreover, Davies himself acknowledges the "many typical features" of Lesley's character's situation even as he grants that a Euro-American reader is not likely to recognize distortions (Davies 1994: 243). Perhaps it is not Lesley's intent to generalize, but the fact remains that most people in the dominating society wherein his work is received know Indigenous North Americans only through the aggregate of such "particularist" representations and this aggregate ultimately forms the basis of their generalized assumptions/depictions/claims. Thus, Lesley's novels are themselves a part of the corpus of constructed images of Indigenous North Americans against which they would be checked by members of the dominating society for relative context.

Taken in this light, Davies's assertion that the rejection of a "human nature" is as ideological a position as its affirmation (a point which, by itself, I would not dispute), besides missing the point, seems somewhat superfluous. It might be argued that Lesley cannot be faulted personally if others take from his work something other than what he intended—that is, in drawing generalized assumptions about the Nez Perce or Indigenous North Americans more generally. But it might just as well be argued that he should reasonably have understood that the required frame of reference by which most members of the dominating society might have recognized the particularist nature of his characters is lacking. In this context, to argue that one bears no responsibility for what others might derive from one's work is itself a political/ideological stance rooted in the tradition of Western liberal individualism. Therefore, to make this argument against Indigenous North American critics of works like Lesley's is, in the end, an incumbent imperialist practice in its own right.

The Violences of Erasure

If there is any lingering doubt about the extent to which the essentialization and objectification of Indigenous North Americans in popular culture has conditioned the nature of the relationship between Indigenous peoples and the dominating society, one need only look to the all-too-common practice of employing distorted and often caricatured Indigenous imagery for everything from commercial trademarks to sports team mascots. In an essay entitled "Let's Spread the 'Fun' Around: The Issue of Sports Team Names and Mascots," Churchill runs through a litany of familiar and profoundly offensive appellations for a number of different racial and ethnic minorities, sarcastically suggesting how each might be applied in the identification of some fictional professional sports franchise (Churchill 1994: 65–72). If Churchill's polemic is offensive and difficult to read—and it is—that is precisely the point. One is hard-pressed, after all, to imagine any other racial or ethnic group that could be treated with such disrespect and be so publicly lampooned without raising in the dominating society a general sense of impropriety.

Even as we acknowledge the relatively longer-term function of figural discourse in support of the advanced colonial system, it is important that we do not lose sight of the more immediate consequences of the deployment of distorted images of Indigenous North Americans. Being exposed to these images themselves, many Indigenous people may internalize elements of them, and this may, in turn, lead to varying degrees of their participation in their own subjugation. One study, for example, which inquired into the racial and cultural identification and preferences of Indigenous children aged 3 to 7 on an Ontario reservation in the early-1970s, found that the children held mostly negative associations toward their own in-group. This led researchers to conclude that "[t]he dominant group's stereotyped negative attitudes and values toward Indians have been communicated to the Indian children through their own families, friends, authority on the reserve, schools, radio and television, etc., and these attitudes and values have been accepted by the children" (Grindstaff, et al. 1973: 377). It is also important to recognize that stereotypes about Indigenous people(s) need not be overtly and explicitly negative to have a similarly detrimental effect upon self-esteem—the superhuman qualities sometimes attributed to the generic "Indian" of popular culture can be every bit as damaging. Dorris notes, for example, that Native children may feel pressure "to live up to their mythic counterparts and feel like failures when they cry at pain or make noise in the woods" (Dorris 1987: 104).

The unequal power relations of advanced colonialism leave open few avenues of resistance to these violences other than those that can be built around questions of authenticity. Noting the problems for Indigenous communities that accompany the dominating society's interest in Indigenous North American culture, W. Roger Buffalohead has insisted that "Indian nations and tribes should be the final arbiters of cultural authenticity and, according to their own needs, should determine cultural directions for the growth and development of their communities" (Buffalohead 1992: 199). However, the overwhelmingly disproportionate capacity of the dominating society to author-ize and disseminate accounts and images of Indigenous peoples makes this seem a faint hope at best. The practices of the past show no sign of abating. On the contrary, Indigenous North American imagery and ideas—however distorted—have been very much in vogue of late.[12] And just as the deployment of constructed images of Indigenous people(s) continues in earnest, we should expect their essentialized, objectified, stereotypically derived content to remain relatively unchanged so long as the power relations that underlie them are intact. With specific reference to a particular medium, Churchill argues that "[t]he dehumanizing aspects of the stereotyping of American Indians in American literature may be seen as an historical requirement of an imperial process" (Churchill 1992: 29). If this is so, as they pertain to Indigenous North Americans, linguistic and figural discourses are inseparable from the status quo of unequal power relations between Indigenous people(s) and the dominating society. As Dorris puts it:

> It is little wonder, therefore, that Indian peoples were perceived not as they were but as they *had* to be, from a European point of view. They were whisked out of the realm of the real and into the land of make-believe. Indians became variably super and subhuman, never ordinary. They dealt in magic, not judgment. They were imagined to be stuck in their past, not guided by its precedents.   (Dorris 1987: 102; emphasis in original)

Moreover, if discourses about the Other are locked in a mutually constitutive relationship with deeply held ideas about the Self, what is at stake in challenging constructed images of Indigenous people(s) may be too much for many members of the dominating, advanced colonial society to bear. As if designed to illustrate this point, a 1991 exhibition at the Smithsonian's National Museum of American Art entitled "The West as America: Reinterpreting Images of the Frontier, 1820–1920" drew considerable criticism from members of Congress, scholars, the press, and the public for the curators' interpretations of Frontier-era art as racist, sexist, and imperialist (Price 1993: 230–31). In the words of one apparently incensed visitor to the museum: "Despite [the curators'] best efforts at 'political correctness,' they have produced a show whose visual impact confirms all that they detest—the expansion westward was good, desirable, and brought the 'New World' into the civilized mainstream" (Smithsonian Institution 1991: 8).[13] In light of such avowals, Campbell's characterization of the United States as "the imagined community *par excellence*" (Campbell 1996: 166) seems extraordinarily well founded.

All such accounts and constructions of Indigenous people(s) help to perfect and sustain an advanced system of colonial domination. Of course, the objection that we live in an era after colonialism will continue to be raised. And it may be pointed out that, at least in North America, the coercive powers of the settler state are no longer brought to bear upon Indigenous people(s) in order to effect their forced subjugation. Leaving aside Mexico as a somewhat special case, we may grant that, in the main, this is true—although, even in Canada and the United States there have been many notable exceptions.[14] It would, in any event, be a mistake to interpret the *relative* absence of direct, physical coercion as evidence of a weak and underdeveloped system of colonial domination. Quite to the contrary, this is suggestive of an advanced and well-established colonial system, inasmuch as the subjugation of Indigenous North Americans has been perfected to such a great degree that the power of ideational constructs and of the everyday structures of inequality—what Foucault calls capillary power[15]—has, for the most part, proved sufficient to sustain and reproduce the system.

The objectification of Indigenous North Americans is central to the maintenance of this system. It is this practice that facilitates the fixing of spatial and/or temporal parameters on Indigenous people(s). This, in turn, fosters the development of essential knowledges about Self and Other that are transmitted through a variety of discursive forms, using the popular cultural media of the dominating society as vehicle. Thus, the rigid spatial parameters of the museum dioramas, the shelf in the souvenir shop, and the cupboard in the movie, inasmuch as they physically contain discursive representations derived from Indigenous North American imagery, may be seen as metaphors for the assignment of place and time imposed upon real Indigenous people(s). Even where this seems not to have been undertaken instrumentally, an apparent predisposition to perceive Indigenous people(s) according to certain deeply entrenched assumptions emerges. This derives, in part, as much from the inadequacy of critical examination of constructions of the Self as of the Other. In the end, it infects even the most benignly conceived representations of Indigenous people(s)—whether as narrator or character, subject or device, in the final analysis all are to some degree like the "Indians" in the museum, on the shelf, and in the cupboard: bounded, objectified, and, ultimately, discursive.

Notes

1. Though I cannot engage debates on the relationship between the discursive and the figural here, I am following Gilles Deleuze's (1988) reading of Foucault in which discourse is not reducible to linguistic forms alone.

2. The process is simultaneous in the sense that presentation and construction are, as in Foucault's formulation, inseparable. It is reciprocal to the extent that it is not a singular act, but an ongoing process of (re)production wherein each presentation reconfirms the essentials of the aggregate construction that, in turn, informs subsequent presentations. Through repeated performance, constructed "knowledges" of the Other are thus disseminated such that audiences judge the veracity of performance (and counter-performance) against what is already "known." This admits of incremental change through cumulative minor variation—images of the Other are forever works in progress—but is remarkably resistant to any dramatic shift in script.

3. It is noteworthy that, as in the film's name, the generic appellation "Indian" is used here, as opposed to the more specific Onondaga identity ascribed to the Little Bear character.

4. Though Jefferson was not the sole author of the Declaration, the reference here to the "merciless Indian Savages" and the account of their "known Rule of Warfare" also appear in a surviving early draft that he is known to have written. This is an especially good example of the centrality of the idea of the savage to the project of Euro-American state-making inasmuch as it is invoked here not in the context of a struggle with Indigenous peoples themselves, but for the purpose of indicting George III.

5. See also Coser (1956: 87–110).

6. The point of this first diorama seems to have been to assert the origins account advanced by Western academics that has it that the first people to arrive in the Americas came from Asia by way of a land bridge across the Bering Sea. That this account contradicts Indigenous peoples' own creation stories is not at all uncontroversial. See, e.g., Churchill (1995: 265–96). The point in noting all of this is not to join the debate about origins but to underscore whose voices can be heard in the Ontario Archaeology Gallery and whose cannot.

7. See, e.g., Boehmer (1995: 20).

8. Recall Campbell's notion of discursive economy.

9. That is not to say that there are not a great many members of the dominating society who would be inclined toward active and enthusiastic support of the colonial project. The point is simply that instrumental design is no more a necessary than a sufficient condition of advanced colonial domination.

10. This, of course, is not to say that life without the trappings of modern Euro-American society is objectively "primitive." However, as rendered, the mannequins corresponded to what members of the dominating society are likely to expect of "Indians" as much as they portrayed lifeways that we know are no longer the norm in Ontario. Consequently, the Indigenous peoples of Ontario are not

accorded a place in the present. Interestingly, the exhibit of European body armor displayed on an upper floor of the museum included a padded leather suit of the sort worn by contemporary motorcycle racers.

11. For an excellent discussion of the processes by which the cultural and intellectual property of Indigenous peoples is first situated in the public domain and then reconverted into the individualized property of persons in the dominating society, see Whitt (1995a).

12. This phenomenon has not been restricted to popular culture. Even in the elite culture's opera circles, there has been a recent revival of interest in ostensibly Indigenous North American themes. See Wynne (1996).

13. As Churchill has observed, "those who read, write and publish American literature are unfamiliar with and quite unwilling to acknowledge their truth as myth; it is insisted upon in most quarters that the myth *is* fact" (Churchill 1992: 19; emphasis in original).

14. Wounded Knee, South Dakota, Oka, Quebec, and Gustafson Lake, British Columbia are but three place names that have become virtually synonymous with contemporary confrontations between Indigenous North Americans and the paramilitary and/or military forces of the colonial state. Indigenous people in Canada and the United States are also imprisoned in numbers far exceeding their demographic proportions in the overall populations of those countries.

15. Recall that for Foucault this signifies the working of power through our ideas, our attitudes, and even the most mundane aspects of our daily lives (Foucault 1980: 39). And even as it works *upon* them, power in this sense also works *through* and is (re)produced by the oppressed themselves (Foucault 1977: 27). See also, Foucault (1978: 92–98).

CHAPTER 6

Travelogues: The Ethnographic Foundations of Orthodox International Theory

The near-complete neglect of Indigenous peoples by International Relations is a loose thread that now deserves a few more purposeful tugs. Critical reflection upon the sources of this lapse gives rise to some important insights into the concealed commitments that underwrite mainstream theory and exert considerable authority in defining and delimiting disciplinary problems, prospects, and possibilities. The origins of these conceptual predispositions and of the neglect of Indigenous peoples can be traced to the travelogues of the first Europeans in the Americas, the enduring influence of which in social contractarian thought recommends their treatment as foundational texts of the social sciences. This view highlights the relevance for International Relations of challenges raised against the veracity of these formative ethnographical accounts inasmuch as such reevaluations simultaneously call into serious question some of the most fundamental ontological commitments of orthodox international theory—commitments that have their conceptual origins in the travelogues. Significantly, the neglect of Indigenous peoples is inseparable from the not inconsiderable conceptual indebtedness of orthodox international theory to these earliest writings about the peoples of the Americas. To the extent that the accounts and claims contained therein are not sustainable in the face of challenges brought against them in recent anthropological literature and cannot be reconciled with autoethnographical accounts of the peoples whose lifeways they purport to document, then,

Realist-inspired International Relations theory becomes identifiable as an advanced colonial practice for perpetuating the erasure to which they give rise.

It might be argued that Indigenous peoples have never constituted a subject matter appropriate to the focus of the field inasmuch as none has ever been possessed of the principal preoccupation of its mainstream scholarship: the Westphalian state. But neither were the ancient Greeks, so that one is left to wonder at the comparatively greater attention devoted to Thucydides' account of the Peloponnesian War and its alleged relevance to the study of contemporary international politics. We may also wonder what marks Indigenous American peoples' statelessness as very much different from that of the Palestinians or the Kurds, both of which groups have been spared the same degree of neglect. And lest the objection be raised that the politics of the Indigenous nations of the Americas have been specific to their places within the states in which they are spatially embedded and have not extended into the "international" realm, it is well to remember that the enactment of treaties has been a widely used instrument in relations between the First Nations and the colonial powers. Similarly, the presence of Indigenous peoples' representatives at the United Nations under the auspices of the Permanent Forum on Indigenous Peoples builds on a history of such involvements that began with attempts to gain standing at the League of Nations and, earlier still, with delegations to the royal courts of Europe.[1] Moreover, Indigenous nations, as we have seen, have their own histories and traditions of sociopolitical organization and inter-national interactions that predate the advent of the European settler state. How, then, do we account for the failure of International Relations scholars to see them?

Darby and Paolini note that International Relations was similarly inattentive to the rest of the non-European world prior to the era of decolonization (Darby and Paolini 1994: 380). Owing to the subsumption of the colonies into the various European empires, their external relations were not understood to be "international." Not until they became intelligible to it by way of the proliferation of statehood in the mid-twentieth century could International Relations engage the former colonies—and this suggests a great deal about the sources of the continued invisibility of peoples in places where decolonization has not occurred. Likely the most important determinants of International Relations' neglect of Indigenous peoples, then, are hegemonic accounts of the possibilities for political order in respect of which the state is treated as monopolistic. The ontological commitments of the theoretical orthodoxy of the field, chief among which is an abiding faith in a Hobbesian-inspired view of human nature, foreclose the possibility of political community in the absence of state authority. Hence, not only are the Indigenous peoples of the Americas rendered invisible to the International Relations orthodoxy, but it also becomes possible to characterize the settler states resident on their territories as *former* colonies, thereby mystifying the contemporary workings of advanced colonialism. In this sense, the undifferentiated idea of the state, making no distinction with respect to settler states, obscures even the fact of the very obscurities it sets in train. This construction turns principally on a prior acceptance of the Westphalian state as the only possible—or, at least, the only legitimate—expression of political order. But if competing claims that imagine a broader range of possibilities cannot be reconciled with the orthodox sense of "reality," it is important to appreciate, as we shall see, that the concepts and categories of the orthodoxy itself are what define the limits of its sense of the "real."[2]

Although, as R.B.J. Walker points out, Hobbes' radical conception of the autonomy and equality of individuals in the state of nature—a condition that is fundamental to the emergence of anarchy—does not lend well to the unequal relations between states (Walker 1993: 93), the derivative idea of an "international anarchy" remains axiomatic to the theoretical orthodoxy of International Relations. Scholarship situated in this tradition is in the same instant furnished with its unit of analysis, the state, and committed to a circumscribed conception of political community, once again, the state. That these commitments undergird the theoretical edifices of the orthodoxy marks out, in turn, a very limited ontological terrain upon which to imagine security, sovereignty, community, and the metaphysics of the good life. Thus, with respect to one of these, Walker argues that "[s]ecurity cannot be understood, or reconceptualized, or reconstructed without paying attention to the constitutive account of the political that has made the prevailing accounts of security seem so plausible" (Walker 1997: 69). Imperiled in any contestation of the appropriateness of the state as the referent object of security, then, are deeply held commitments with regard to the possibilities of political order—possibilities that are presumed to begin and end with the state.[3] Remarkably, this whole assemblage of convictions rests upon an unsubstantiated idea: the anarchical state of nature.

The Western philosophical origins of the order/anarchy dichotomy reside most famously with Hobbes and the other social contractarian theorists of the so-called Age of Enlightenment. But though they advanced and elaborated their ontological commitments with airs of certain knowledge and experience, most of these theorists had never seen the "natural" worlds so fundamental to their philosophies. Rather, they relied on accounts from travelogues authored by persons on the leading edge of the European empires' march into the rest of the world. This has led Peter Hulme and Ludmilla Jordanova to suggest that these lesser-known writings from the frontiers of European imperial expansion ought to be considered as Enlightenment texts. Following from this proposition, they find the canons of social contractarian thought implicated in the imperialist project: "Some of the principal works of writers like Hobbes, Locke, Rousseau, Ferguson, and Diderot draw on accounts of these travellers in ways both important for their status as key texts of the Enlightenment, and revealing of their implication with the whole process of European exploration and colonization of the non-European world" (Hulme and Jordanova 1990: 8). Given this connection and the enduring influences of social contractarian thought, we should also regard the travel writings as foundational texts of the contemporary Western social sciences. Though we do not read these accounts directly in International Relations, they insinuate themselves through underinterrogated ontological commitments of mainstream Realist-inspired international theory, most conspicuously in hegemonic renderings of the state of nature. Here, then, are the unacknowledged ethnographies that inform orthodox International Relations scholarship.

Interestingly, many of the same assumptions that underpin the orthodoxy of International Relations and its more fundamental political commitments may be found at the root of a number of orthodox anthropological and historiographical accounts about Indigenous peoples that cast their pre-Columbian condition in terms consistent with a Hobbesian state of nature. It is therefore instructive to consider some of these accounts and to assess both the integrity of the evidence upon which they rest and the extent to which they can or cannot be reconciled with the traditional worldviews and lifeways of the peoples to whom they ostensibly refer. Such an interdisciplinary approach has much to recommend it inasmuch as anthropologists and historians have been among the most attentive to those other foundational texts of the social sciences: travelogues. This makes it possible to challenge key ontological commitments of the orthodoxy of international theory at their points of conceptual origin.

What is immediately apparent is that the orthodox representations of Indigenous peoples are simultaneously highly gendered and racialized. In this regard too, they share discursive terrain with the orthodoxy of International Relations—as J. Ann Tickner reminds us, "nonwhites and tropical countries are often depicted as irrational, emotional, and unstable, characteristics that are also attributed to women" (Tickner 1992: 7). This has extended even to the discourses of science where, according to Ania Loomba:

> Dominant scientific ideologies about race and gender have historically propped each other up. In the mid-nineteenth century, the new science of

anthropometry pronounced Caucasian women to be closer to Africans than white men were, and supposedly female traits were used to describe "the lower races."   (Loomba 1998: 63–64)

Indeed, the culture/nature dichotomy has always privileged the masculinized European Self as against the exotic feminine Other. And, as Peterson so convincingly argues, this, in turn, has been inseparable from the process of state-making (Peterson 1992). As such, the advent of the settler state is the concrete expression of the parallel centrality of gendered and racialized discourses to advanced colonialism. However, it is important to point out that the discursive renderings of Indigenous peoples also evince a certain ambivalence in both regards: orthodox accounts of the aboriginal condition of Indigenous peoples are permeated with discourses of gender and race, but these discourses are oft times conflicted in themselves. In ways that seem more than coincidentally to befit particular colonial purposes (Henderson 2000: 27–29), Indigenous people(s) are variously constructed as frighteningly masculine or piteously feminine, as supremely rational or hopelessly irrational, as coldly stoic or wildly emotional, as superhuman or subhuman. As with the parallel constructions in pop-culture treatments of Indigenous people(s), this highlights both the contingency of these discourses and the colonial purposes they serve(d). It also draws our attention once more to the important role of negative definition—as expressed through these dichotomies—in the construction of Western self-knowledges.[4] This productive dimension must always be borne in mind along with Foucault's conception of the capillary form of power lest, as Steve Clark cautions, "the power to generate meaning" is otherwise "reduced to cynical legitimation of more basic and brutal mechanisms of power" (Clark 1999: 9).

In the balance of this chapter, the idea that the aboriginal condition of Indigenous peoples is unproblematically apprehendable by way of reference to European accounts from the early contact period is challenged. Evidence of the distortive influences set forth by the arrival of Europeans in the Americas is briefly considered in reference to the Yanomami and Cherokee peoples. The evidentiary bases of several orthodox anthropological and historiographical accounts of the pre-Columbian warfare of the Lakota are then assessed and shown to be reconcilable to a range of conclusions other than the Hobbesian-inspired ones of which they have been deemed supportive. Inasmuch as these same accounts and the hegemonic conceptions of order/anarchy with which they are mutually constitutive render Indigenous peoples invisible to the orthodoxy of International Relations, their indeterminacy suggests that they should rightly be viewed as advanced colonial practices; all the more so when we consider that Lakota traditionalism's non-state articulations of political order and community are simultaneously invalidated by them.[5]

In more concrete terms, this might be little more than academic were it not for the fact that it is an existing community—not an historical curiosity—that is thus marginalized. It should be reemphasized that Lakota traditionalism is by no means a reference to the past or the ethnographic present. Lakota traditionalists make up a sizeable proportion of the *contemporary* Lakota *Oyate* and are characterized most fundamentally by their enduring fidelity to traditional cosmological commitments—a fidelity that has thus far survived the assimilationist practices of colonialism and is significant notwithstanding that some of its referents might turn out not to be wholly unaltered survivances of some precolonial era. As we have seen, a lingering legacy of colonialism is the fragmentation of many Indigenous peoples that tends to manifest most conspicuously in the cleavage between "traditionalists" and the more assimilated "progressives." The Lakota certainly have not been an exception in this regard and have, in fact, suffered some of the worst of the political violence that sometimes attends this division. Thus, while it is important that we remember that they are not representative of the whole of the Lakota people, the traditionalists do constitute a contemporary community that resides quite decidedly beyond the pale of orthodox International Relations theory. And it is precisely the sort of wholesale invalidations of the cosmologically based worldview and lifeways of the traditionalists, in which orthodox International Relations theorists are implicated, which constitute the ideational dimension of the advanced colonial domination to which they are subject. These same conceptual predispositions (among which the Hobbesian impulse is prominent) seem to render traditional accounts of Lakota lifeways quite implausible, thereby undermining their utility as referents for the organization of resistance.

Finally, it should be pointed out that the whole of the Realist tradition in International Relations is touched by the critique that follows. While it is Classical Realism that is most readily associated with Hans Morgenthau's recourse to "objective laws that have their roots in human nature" (Morgenthau 1985: 4), Neorealism does not effect a convincing break from this. Of course, Neorealists have disavowed any recourse to a Hobbesian-inspired account of human nature and, by and large, the field has taken them at their word, accepting the foundational abstraction to an anarchical international *system* of egoistic *states*.[6] But in ontologizing the state instead of human nature, Neorealists fail to give sufficient content to their conception of international anarchy—that is to say, anarchy by itself is the cause of nothing. Structural anarchy might, for example, allow wars to occur, but it does not tell us why they occur; for this, we must know something about those doing the warring.[7] Failing this, we are left without an account of why conditions of anarchy should result in competition and not cooperation.[8] In order to overcome this problem, the ontologized state must also be anthropomorphized through the ascription of self-interest, and this necessarily reintroduces the Hobbesian-inspired commitments of Classical Realism. Robert Gilpin is thus compelled to acknowledge that the "interests" of the state are ultimately those of real people (Gilpin 1984: 301).[9] To speak of the orthodoxy in International Relations, then, is to evoke the whole of the Realist tradition.

Indigenous Peoples and the State(ments) of Nature

Michael Dorris has observed that learning about and from Indigenous North American cultures and histories is rather different from acquiring knowledge in other fields because the researcher more than likely has already received and internalized a huge amount of misinformation about Indigenous peoples that threatens to subvert inquiry from the outset (Dorris 1987: 103). As we have seen, pop-culture has been a powerful and persistent vehicle for such misinformation, and one that has reproduced particular images and accounts of Indigenous peoples by responding to and reconfirming what most members of the dominating society already "know" about them. Just as these images and accounts are thus their own arbiters of authenticity, so too scholarly inquiry is susceptible of being corrupted by what is already "known." That is to say, there is often a great deal that must be unlearned before serious and productive investigation can begin. This is perhaps nowhere better illustrated than in the corpus of literature purporting to elucidate the functions and conduct of warfare in pre-Columbian Indigenous societies. Despite the epistemological predisposition on the part of some scholars working in Anthropology, History, and other disciplines to present their conclusions as matters of objective fact, backed up by the rigors of Western "science," discerning the aboriginal condition of Indigenous peoples is not at all a straightforward and unproblematic undertaking. As Dorris points out, "[i]t depends on the imperfect evidence of archaeology; the barely-disguised, self-focused testimony of traders, missionaries, and soldiers, all of whom had their own axes to grind and viewed native peoples through a narrow scope; and, last and most suspect of all, common sense" (Dorris 1987: 104). Significantly, traditional Indigenous sources are seldom ever consulted, their exclusion typically justified on the grounds that the oral literatures characteristic of so many Indigenous societies are less reliable than written forms. Consequently, the body of scholarship on the histories of Indigenous peoples has been largely self-referential, continually reproducing whatever errors of perception and assumption may derive, per Dorris's reproof, from the application of a generally ethnocentric "common sense."

Convincingly demonstrating this point is an article by renowned military historian John Keegan that follows from his investigations into the history of warfare on the Northern Great Plains of North America, and in which he seems not to have consulted, much less taken seriously, Indigenous sources.[10] One does, however, sense a reliance on a decidedly Western "common sense" in this analysis of the putative facts of warfare on the Plains. Central to this widely accredited wisdom is the familiar Hobbesian impulse that, finding in the aboriginal condition nothing akin to the state as a means by which political order might be furnished, posits a perpetual state of war and insecurity in its stead. Here Keegan is in distinguished company: Hobbes himself maintained as evidence of the plausibility of his idea of the state of nature that "the savage people in many places of *America* . . . live at this day in that brutish manner, as I said before" (Hobbes 1968: 187). While Keegan does not explicitly articulate this assumption, it seems implicit in, for example, his assertion that Custer and his 7th Cavalry were "wiped away in an outburst of native American ferocity" while their intended Lakota and Cheyenne victims are described as having been motivated less by the pressing need to defend their encampment from the attacking soldiers than by their own "ferocious emotions" (Keegan 1996: 41).

Similarly, and perhaps partly in consequence of a prior assumption of unrestrained savagery, Keegan ascribes an entrenched and pervasive individualism to the people of the Plains. Indeed, the Hobbesian overtones of his work are complemented by his characterization of the lifeways of the Plains people(s) as "rigorously masculine and individualistic" (Keegan 1996: 15). Keegan attempts to back up this position by reducing the Sun Dance—a protracted ceremony in which individuals undergo considerable personal suffering as a mode of self-sacrifice on behalf of the whole of their people and as a means by which to gain spiritual enlightenment—to a contest between participants motivated by nothing more than the selfish desire by each to "demonstrate in public his powers of endurance" (Keegan 1996: 15). According to Howard Harrod, "sun dances and other ritual processes provided occasions for individuals to endure the suffering that was requisite for religious experience" (Harrod 1995: 26). Keegan, however, sees, as the only functional outcome of this sacred ritual, the participants' acquisition of "qualities of physical hardness, contempt for pain and privation, and disregard of danger to life that both disgusted and awed the white soldiers who fought them" (Keegan 1996: 15). In this highly racialized discursive construction—in opposition to "awed . . . white soldiers"—the Plains people(s) are rendered as unreal, constructed at what might be termed the super-subhuman nexus.

Keegan is by no means alone in citing individualized motives as the basis of Indigenous peoples' warfare. Anthony McGinnis shares this perspective, arguing that "[i]n war, the tribe was important only insofar as it supported the individual warrior and his combat and in the fact that the tribe's non-combatants . . . needed to be defended" (McGinnis 1990: 12). Emphasizing this point, he draws a comparison to a French officer, Pierre de la Verendrye, who was wounded at the Battle of Malplaquet in 1709: "Fortunate enough to recover from his wounds, Verendrye returned home to Canada, having willingly shed his blood for God, King Louis XIV, and France, something the Indians of the northern plains would not have understood—sacrifice for an ideal or a leader rather than for oneself" (McGinnis 1990: 4). Individuals in Plains societies, according to McGinnis, were prompted into warfare only in order to obtain wealth and glory for themselves (McGinnis 1990: x). Similarly, John C. Ewers identifies opportunities for individuals "to distinguish themselves" and the pursuit of "coveted war honors" as important determinants of warfare between Plains peoples (Ewers 1975: 401). The hyper-individualism in these accounts dehumanizes to the extent that it precludes all but the barest sketches of a social world; for lacking in loyalties more profound than the satisfaction of their personal appetites, Indigenous people are thus rendered all the more frightening.[11]

But perhaps the most extreme position as regards the presumed individualized sources of aboriginal warfare is advanced by Napoleon A. Chagnon. Chagnon's account of warfare among the Yanomami people of Amazonia finds biological determinants prominent among its sources. Central to his argument is the idea that Yanomami warfare, though sustained by a revenge complex wherein violence by one group reciprocally begets violence in kind from its erstwhile victims, is, at base, motivated both by competition over scarce material resources and by a supposed biological imperative on the part of males in kinship-based groups to secure, by means of violence if necessary, enhanced access to "reproductive resources"—that is, women. According to Chagnon:

> It is to be expected that individuals (or groups of closely related individuals) will attempt to appropriate both material and reproductive resources from neighbors whenever the probable costs are less than the benefits. While conflicts thus initiated need not take violent forms, they might be expected to do so when violence on average advances individual interests. I do not assume that humans consciously strive to increase or maximize their inclusive fitness, but I do assume that humans strive for goals that their cultural traditions deem as valued and esteemed. In many societies, achieving cultural success appears to lead to biological (genetic) success. (Chagnon 1988: 985)[12]

This formulation clearly hints at what sociobiologists have termed the "selfish-gene" concept: the idea that certain social behaviors are, at least in part, biologically determined and that the resultant social outcomes are a determining factor in the evolutionary natural selection of species. Put another way, it presumes to "show that there are evolutionary 'optima' for behaviours such as aggression" (Van Der Dennen and Falger 1990: 15). But what may be most interesting about this argument from the point of view of someone who works primarily in the field of International Relations are the similarities it shares, in several particulars, with Realism. There is, of course, the obvious implication that human nature—or at least the nature of the "successful," in evolutionary terms—is as Hobbes imagined. And absent the state, it is individuals—more particularly, individual men—who are cast as the "rational gains maximizers," such that the possibility of political order is effectively precluded. Having thus found his subjects residing in a Hobbesian state of nature, Chagnon, like Keegan, McGinnis, and Ewers, sets about explaining the sources and conduct of their wars in terms consistent with this condition.

If Chagnon is right and warfare in Amazonia is indeed in some significant measure a function of genetic fitness, then it would logically follow that the apparently warlike tendencies of the Yanomami can, with confidence, be mapped back onto their pre-Columbian ancestors. Furthermore, if this behavior is biologically determined, it must be specified as a general human characteristic. The imposition of Hobbes' Leviathan, then, would serve to explain why it is that the conduct of the Yanomami is peculiar and not universal to the human condition. The political implications of such an inference are simultaneously abstract and immediate: in the abstract sense, it would seem to lend support to the notion of the state as the sole locus of political order; more immediately, it confers moral approbation upon the conquest of Indigenous peoples and the suppression of their traditional lifeways, if only (at least ostensibly) to save them from themselves. Indeed, as Jacques Lizot points out, Brazilian newspapers supporting the interests of resource industries that have been accused of orchestrating genocide against the Yanomami in order to gain access to their lands have enthusiastically embraced Chagnon's writings (Lizot 1994: 845).[13]

Here again, the Hobbesian impulse is not anomalous. It is as readily invoked as a justification for past conquests as for those that are ongoing. Though he does not follow Chagnon onto the thin ice of sociobiology, Ewers (apparently oblivious to the sum and substance of Dorris's warning about the questionable reliability of early Euro-American sources) argues that "intertribal warfare was rife [on the Northern Plains] at the time these Indians first

became known to whites" and that this "is evident in the writings of the pioneer explorers" (Ewers 1975: 399). Although he acknowledges that there is scant evidence that is suggestive of large-scale battles, presumably with the aim of demonstrating that the possibility of large-scale exterminative warfare was not precluded, Ewers cites the example of an 1866 battle in which "the Piegan are reputed to have killed more than three hundred Crow and Gros Ventres" (Ewers 1975: 401). Nevertheless, inasmuch as raiding for horses was the principal form of warfare among Plains peoples, he submits that this is likely to have been the primary source of casualties (Ewers 1975: 402). With an apologist agenda beginning to show, he continues: "Nor is there reason to doubt that, during the historic period, many more Indians of this region were killed by other Indians in intertribal wars than by white soldiers or civilians in more fully documented Indian-white warfare" (Ewers 1975: 402). Having thus outlined the rudiments of a Hobbesian state of nature as extant on the Northern Plains at the earliest stages of European contact, Ewers makes what seems a thinly veiled attempt to rationalize the forced imposition of the Euro-American Leviathan, proposing that "[h]ad each of the tribes of this region continued to stand alone, fighting all neighboring tribes, it is probable that many of the smaller tribes either would have been exterminated, or their few survivors would have been adopted into the larger tribes, thereby increasing the latters' military potential" (Ewers 1975: 402).

Once more, then, the aboriginal condition has been presented as representative of a state of nature constructed in decidedly Hobbesian terms, here echoing nineteenth-century colonial tropes on the presumed inevitable extinction of "savage" peoples.[14] But what Ewers seems to miss is the possibility that the aboriginal condition of the peoples he studies is not, in fact, known to him. In this too he keeps company with Keegan, McGinnis, and Chagnon. In contrast, R. Brian Ferguson raises a compelling challenge to the pretension of scholars such as these to know the pre-Columbian lifeways of the Indigenous peoples of the Americas, regardless of whether their focus is on the conquered and colonized Plains peoples of North America or the as yet largely unsubdued Yanomami of Amazonia. Investigations by Ferguson in which he focused primarily on the Yanomami suggest that, contra the received wisdom of the Hobbesian impulse, "the most general cause of known warfare in Amazonia is Western contact" (Ferguson 1990: 237). Although he does not contend that warfare was unknown to pre-Columbian Amazonia, he does insist that, "[c]ontrary to Hobbes, the intrusion of the Leviathan of the European state did not suppress a 'war of all against all' among Native peoples of Amazonia, but instead fomented warfare" (Ferguson 1990: 238). "Ultimately," he continues, "wars have ended through pacification or extinction, but prior to that the general effect of contact has been just the opposite: to intensify or engender warfare" (Ferguson 1990: 239). Moreover, Ferguson holds this to be a general consequence of European imperialism virtually wherever it has confronted non-state societies, albeit with notable local variations arising from indigenous peculiarities (Ferguson 1992: 109).

Ferguson (1990: 239) attributes this phenomenon to an array of influences that fall roughly into three broad categories. The first is concerned with the purposeful incitement and/or direction of Indigenous warfare by Europeans. Such practices were very common in the initial contact period and were manifest in a variety of forms. The most obvious and direct of these was the use of conquered or allied Indigenous peoples as "auxiliaries or impressed recruits" in European campaigns against unsubjugated peoples on the peripheries of the expanding colonies (Ferguson 1990: 239). In some cases, notably along the line of confrontation between the English and French colonies of northeastern North America, Indigenous peoples were unable to avoid becoming entangled in wars between the colonial powers themselves. Elsewhere, Europeans found it expedient to facilitate—generally by the provision of arms and other goods—warfare amongst contending groups lying beyond the reach of direct colonial authority. And Tom Holm notes that the militarization of Indigenous American peoples also served European interests as "a method of absorbing them into a larger imperial system" (Holm 1997: 462).[15] Ewers, however, rejects the idea that European contact incited warfare between Indigenous peoples in this way and, as evidence, points to the matter of the support that was given by Euro-Americans to the Crow and Arikara in their struggles against the Lakota:

> To view the Crow and Arikara as "mercenaries" of the whites is to overlook the long history of Indian-Indian warfare in this region. The Crow, Arikara, and other tribes had been fighting the Sioux for generations before they received any effective aid from the whites. They still suffered from Sioux aggression during the 1860s and 1870s. Surely the history of Indian-white warfare on the northern Great Plains cannot be understood without an awareness of the history of intertribal warfare in this region. (Ewers 1975: 409–10)

But, while Ewers is right in pointing out that the colonial powers, by means of exploiting existing animosities between some peoples, frequently did not need to rely on coercion in enlisting the service of Indigenous recruits, in the end it was still these powers that enkindled enmity into open hostilities.[16]

A further impetus toward the deliberate and utilitarian incitement and direction of warfare between different Indigenous peoples was the European demand for slaves in the early stages of the colonization of the Americas. As Ferguson explains it:

> The initial European colonization of the New World was based on the coerced labor of Native peoples. Adult male captives were sought as field laborers, women and children as domestic servants. Royal decrees—which were often circumvented but which still had an impact—allowed two main avenues for enslaving Indians: taking captives in "just wars" against allegedly rebellious Natives or putative cannibals; and "ransoming" captives held by Indians from their own wars. It was the latter that became the routine source of slaves. . . . Slaving was encouraged by payments in European goods, but raiding was not entirely optional; people who did not produce captives were commonly taken as slaves themselves. Slave raiding was often a constant danger even hundreds of miles from European settlements.   (Ferguson 1990: 240)

Wilma Dunaway draws to our attention similar conditions that had a profound effect upon the nature and extent of warfare as practiced by the Cherokee during the early period of contact in southeastern North America:

> Prior to the development of a profitable market for war captives, slaves remained only a by-product of conflicts waged primarily for vengeance. Cherokee clans frequently adopted prisoners of war to replace kinsmen who had died, or captives could be ransomed by the enemies. Once the traders began exchanging goods for war captives, the market value of the captured slaves intensified the frequency and extent of indigenous warfare. (Dunaway 1996: 462)

Thus, peoples who may never before have been enemies, or perhaps had never even come into direct contact with one another, developed enduring mutual malevolence.

Ferguson's second broad category is concerned with demographic pressures arising from European colonization and the influences they exerted on warfare as conducted by Indigenous peoples. The introduction of epidemic diseases against which Indigenous people had little or no immunity was, according to Ferguson, a source of increased hostility between groups when it led to charges of sorcery (Ferguson 1990: 241). In some instances, catastrophically high rates of mortality due to disease spurred raiding with the express purpose of acquiring captives to be integrated into the abductors' society as a means of population replenishment (Ferguson and Whitehead 1992: 9). Of greater consequence, however, were the migrations prompted by epidemics, slave raiding, and the ever-expanding colonies themselves. Migration forced direct contact between historically separated groups and increasingly brought them into conflict as refugee groups sought to impose themselves into regions that were already well populated (Ferguson 1990: 242).

The third and final set of transformative influences identified by Ferguson is associated with the introduction of Western manufactures. Owing to the greater efficiency of steel tools and other Western goods, such as firearms, vis-à-vis their indigenous equivalents, European trade wares dramatically increased the war-making potential of many Indigenous peoples. These items thus became both objects and implements of war with the deleterious effect that warfare became a means by which to forcibly appropriate the instruments of warfare that, in turn, made possible its expansion and the appropriation of still more of its instruments. It is almost certainly more than mere coincidence, then, that Indigenous peoples who enjoyed ready access to these goods are frequently the same ones regarded as most warlike in Euro-American ethnographies and historiographies. Jeffrey Blick, for example, notes that the gun-toting mounted warriors of the Plains owed their reputation as a warlike people largely to the historical accident of having been situated at the point at which the lines of trade in firearms supplied by the French in the northeast of the continent first intersected with the diffusion of horses introduced by the Spanish in the southwest. As Blick puts it:

> The combination of the gun and the horse . . . enabled many tribes to expand their traditional ranges and to wage warfare in a much more efficient manner. What ultimately resulted was an unequal access to guns and horses. Tribes of the Great Plains proper were able to take advantage of the geographic continuity of the Plains and of the rapid diffusion of the horse and gun. Marginal tribes however, such as the Bannock and the "Digger" Indians of the Plateau and Great Basin, were forced to retreat into inhospitable regions to avoid the raids of their mounted predators, the Blackfoot, Piegan, Shoshone, etc.   (Blick 1988: 666–67)

Thus, we see here the confluence of two broad sets of influences as the migratory pressures felt by the Plains peoples in the face of the advancing Euro-American colonies, combined with their acquisition of horses and firearms, induced warfare with other Indigenous peoples, thereby setting in motion still more waves of migration with all of the disruptive effects that entailed.

It must be emphasized that Blick's position, like Ferguson's with respect to the Yanomami, is not that warfare was nonexistent on the Plains before the introduction of Western manufactures, but rather, that the appearance of these items was typically accompanied or followed in short order by an increase in the frequency and intensity of warfare.

All of this makes Ewers's earlier-cited admonition to take into account the history of intertribal warfare on the Plains seem rather more problematic than it might at first appear. It also serves to underscore Dorris's suggestion that the early Euro-American accounts of the aboriginal condition of Indigenous peoples may be unreliable—a point that he is not alone in making.[17] Ferguson echoes Dorris's concerns, arguing that the first accounts of contact with Indigenous peoples tended to come from "the most disruptive observers imaginable: raiders seeking slaves or mission 'converts' " (Ferguson 1990: 238). Moreover, he poses as a more general problem for Anthropology itself the fact that the first literate observers are seldom present at the time of initial contact:

> [E]thnology is built upon a paradox. Traditionally, it has sought the Pristine Non—non-Western, nonliterate, noncapitalist, nonstate. Yet the quality of our descriptions of other cultures is generally in direct proportion to the intensity of the Western presence. Literate observers usually arrive rather late in the encounter. The specter haunting anthropology is that culture patterns taken to be pristine may actually have been transformed by Western contact. (Ferguson 1990: 238)

But, setting aside for the moment the issue of veracity and the question of timeliness, an even more serious problem from the point of view of anyone hoping to access the aboriginal condition of Indigenous peoples through the accounts of observers, whether contemporary or historical, is the fact that European influences have repeatedly preceded Europeans themselves, changing the lived realities of Indigenous peoples long before first contact. This problem effectively precludes reliance on the accounts of observers with respect to the "pristine" condition of aboriginal warfare: refugee migrations, almost by definition, precede the advance of colonial frontiers; following indigenous trade routes, manufactured goods can become commonplace in a given locale centuries before first contact; epidemic diseases are borne by refugee flows as well as along trade routes. By way of example, the winter counts of the peoples of the Northern Plains indicate a very high frequency of epidemics dating back to 1714, with the first recorded outbreak among the Oglala Lakota having taken place in 1780[18]—24 years before they were first visited by the renowned Euro-American explorers Meriwether Lewis and William Clark in 1804.

So, whether our focus is on the Yanomami of Amazonia, the Cherokee of southeastern North America, or the Lakota of the Northern Plains, accounts of the supposed aboriginal condition of Indigenous peoples that rely to any significant extent upon what was, or may yet be, empirically observable are highly suspect. How, then, do we proceed? Douglas Bamforth proposes that, if "ethnohistoric documentation of warfare tells us little about precontact circumstances," this leaves "archaeological data central to any understanding of post-contact changes in these circumstances" (Bamforth 1994: 97). Accordingly, he directs us to consider the evidence uncovered in the excavation of agriculturally based pre-Columbian settlement sites along the Missouri Trench in present-day North and South Dakota, with particular emphasis on one site at Crow Creek. As a control case, he also discusses the Larson site, an excavation of a large former Arikara community near the Missouri River that was occupied between 1750 and 1785, by which time the disruptive influences of the arrival of Europeans on the continent should certainly have been keenly felt. Bamforth notes that trenches and palisades were generally common features of all of these sites, though the extent of their overall development and completeness as well as the degree of attention paid to their maintenance varied across time (Bamforth 1994: 106). Bamforth, probably accurately, interprets these features as defensive fortifications. But this assumption, in part, leads him to another rather more tenuous one: namely that large-scale exterminative warfare was "endemic" on the Northern Plains even prior to the arrival of Europeans on the continent.

Bamforth bases this position primarily on evidence uncovered in the excavations of the Larson and Crow Creek sites. The latter town is estimated, according to Bamforth, to have been built sometime in the early part of the fourteenth century (Bamforth 1994: 106). It was at this site that a particularly grisly discovery was made in 1978: a mass grave in which were interred the skeletal remains of somewhere in the neighborhood of 500 people.[19] In addition to the fact of their having been buried together in a mass grave, the condition of the human remains at Crow Creek indicates that the inhabitants of the town almost certainly were the victims of a massacre. A very high frequency of depressed fractures to the skulls of the victims as well as other similar indications would seem to make at least this much irrefutable. Significantly, analysis of the skeletal remains yielded a further insight into the tragic situation of the victims: telltale signs in the condition of many of the long bones indicate that the townspeople had suffered from malnutrition at various points in their lives and many of them were malnourished at the time of the massacre

(Bamforth 1994: 106–07). According to Larry Zimmerman and Lawrence Bradley, "[a]ctive and organizing subperiostial hematomas along with the other bony alterations provide convincing evidence that nutritional deprivation had been present for some time prior to the deaths of these people and probably was rampant at the time of their demise" (Zimmerman and Bradley 1993: 218). This, then, suggests a motive and context for the slaughter: forcible appropriation of foodstuffs during a famine. Bamforth compares this evidence to that found at the post-contact Larson site where a similar massacre took place approximately four and one half centuries later, likely in consequence, he argues, of the conflict created by mass migrations that were, in turn, a result of the same disruptive influences of European colonialism identified by Ferguson and Blick (Bamforth 1994: 101–02). And finding the same sorts of osteological evidence—with the exception that indications of malnutrition were not found at the Larson site—and similar fortifications at the two sites, he arrives at the conclusion that "precontact tribal warfare on the northern Great Plains resulted from indigenous cultural-ecological processes rather than from external influences" (Bamforth 1994: 109).

As noted earlier, Bamforth is probably right in regarding the ditches and palisades of the villages in the Missouri Trench as defensive fortifications. Less clear, however, is the conclusion that these measures were undertaken in response to endemic large-scale warfare in the region as a feature of its various peoples' aboriginal condition. Yet, this is precisely what Bamforth implies when he suggests that the construction of such defenses would have been a tremendous burden for such small populations (Bamforth 1994: 111). To be fair, he does acknowledge that "features which archaeologists interpret as fortifications could have primarily symbolic or ceremonial significance . . . or . . . could have served simply as warnings which by themselves dissuaded rival groups from resorting to all-out war" (Bamforth 1994: 105). Ewers, on the other hand, is considerably less cautious: "Surely the prehistoric villagers would not have taken elaborate steps to fortify their settlements had they not been endangered by enemies" (Ewers 1975: 399). And, "[w]hoever those enemies were," he continues, "we can be sure that they were other Indians" (Ewers 1975: 399). But can we, in fact, be so sure of all of this? What if the fortifications—if, indeed, they have been correctly interpreted as such—were inspired by a fear of attack rather than the experience of it? The very fact that, at least in the cases of the Larson and Crow Creek sites, they would seem to have been unequal to the purpose ascribed to them, suggests the possibility that they were designed in response to some lesser threat. In this regard, it is significant that the Northern Plains was noted for small-scale raiding between groups and, especially if archaeologists are correct in assessing periods of food shortage, sedentary agricultural communities, such as the one uncovered at Crow Creek, would have been likely targets of such incursions. Moreover, particularly if we accept Patricia Albers's suggestion that raiding, as a "mechanism for resolving short-term imbalances in the distribution of goods," was a way of maintaining symbiosis between groups (Albers 1993: 108), the complete destruction of a food-producing village would seem contrary to the interests of the raiders and, therefore, unlikely to have been a common enterprise. Some support for this view resides in Bamforth's own observation that the fortifications at Crow Creek had not always been well-maintained as well as in evidence that the village had grown beyond the confines of the encircling ditch that had itself been abandoned and converted to a refuse dump.[20] Of course, none of this is intended to suggest that any of these explanations necessarily represent more accurate portrayals of the reality of pre-Columbian existence on the Northern Plains than those proposed by Bamforth and Ewers. On the contrary, the point here is only to make clear that the archaeological evidence cannot speak to us as unproblematically as Ewers and, to a lesser degree, Bamforth would have us believe.

Bamforth's argument leaves room for a range of conclusions other than those at which he arrives. We may note, for instance, that while he is able to draw our attention to a number of sites along the Missouri Trench, just two bear evidence of large-scale exterminative warfare, and only one of these dates to pre-Columbian times. He does indicate two additional sites at which partially constructed settlements appear to have been abandoned before completion (Bamforth 1994: 105), but his interpretation of this as evidence that the would-be inhabitants had been driven off by force, though a plausible enough explanation, is hardly conclusive. Bamforth acknowledges that the data he examines are more suited to determining the scale of warfare than its frequency, even as he concedes that the fortification of settlements became more common after the arrival of Europeans (Bamforth 1994: 111).[21] One wonders, then, on what basis the Crow Creek massacre should be regarded as anything more than an aberration under conditions that, like the influences set forth from European colonization, were disruptive of the customary lifeways of the peoples concerned. Finally, Bamforth himself draws attention to evidence of famine at the time of the Crow Creek massacre as well as episodically in the years prior. Surely this must be regarded as an extreme circumstance that, though it may well have resulted in a massacre, is in no way indicative of a general trend. In fact, the evidence cited by Bamforth is telling inasmuch as the earlier periods of malnutrition that are also indicated did not result in a similarly catastrophic conflict.

Still, Bamforth and Ewers are not alone in drawing the conclusions they do from the archaeological record. Lawrence Keeley, for example, refers us to the evidence uncovered at the Crow Creek and Larson sites in the course of his direct rejection of Ferguson's thesis (Keeley 1996: 68–69). According to Keeley:

> From North America at least, archaeological evidence reveals precisely the same pattern recorded ethnographically for tribal peoples the world over of frequent deadly raids and occasional horrific massacres. This was an indigenous and "native" pattern long before contact with Europeans complicated the situation. When the sailing ship released them from their own continent, Europeans brought many new ills and evils to the non-Western world, but neither war nor its worst features were among these novelties. (Keeley 1996: 69)

Apart from his somewhat unfair treatment of Ferguson's argument,[22] Keeley lacks a reflexive sense of the ambiguity of the archaeological evidence he cites. And his is also perhaps the most direct example of a Hobbesian-inspired perspective on the aboriginal condition of Indigenous peoples. Concerned at what he regards as "pacified" renditions of the human past, Keeley's purpose is to discredit what is in his view their underlying "theoretical stance that amounts to a Rousseauian declaration of universal prehistoric peace" (Keeley 1996: 20). Accordingly, he appeals directly to Hobbes in support of his argument that, "[i]f anything, peace was a scarcer commodity for members of bands, tribes, and chiefdoms than for the average citizen of a civilized state" (Keeley 1996: 39). And in so doing, he furnishes a clear illustration of the shared ontological commitments underlying both orthodox interpretations of archaeological evidence and Realist-inspired International Relations theory.

The Tyranny of Orthodox Social Theory

Fabian has observed that, "[i]n ethnography as we know it, the Other is displayed, and therefore contained, as an object of representation; the Other's voice, demands, teachings are usually absent from our theorizing" (Fabian 1990: 771). In his study of military patterns on the Plains, Frank Secoy begins by noting that his work is temporally constrained by "the period of the earliest adequate documentary sources for the area" (Secoy 1966: 1). Elsewhere, he tells us that, "[a]nalysis of the Pre-gun–Pre-horse military technique pattern of [the Sioux] must of necessity be incomplete, since there is little factual material on the Sioux during this period" (Secoy 1966: 65). Secoy thus seems to share in the widespread reluctance, exhibited most especially by scholars wedded to objectivist epistemological commitments, to consider the oral literatures of Indigenous peoples as viable documentary sources—a bias that, while reflecting the various prejudices inherent in the culture/nature dichotomy, contributes to the exclusion of Indigenous knowledges and, by extension, of the voices of Indigenous people as well. But, as we have seen, the objection that oral literatures are suspect for being impermanent and susceptible of being altered to reflect the subjective inclinations of their human repositories is not so sound as might be imagined—recall Neta Crawford's observation that "written 'primary' texts are no more omniscient than oral histories" and, in fact, might actually be less so inasmuch as they are not collectively author-ized and more readily mystify their own silences and the biases of their contexts (Crawford 1994: 351). Travelogues and the canons of Western philosophy are, to the orthodoxies of Anthropology and International Relations respectively, "primary" texts of the sort mentioned by Crawford. As much—if not more so—than oral literatures, they reflect many of our deepest and least interrogated assumptions about the world; they reflect, in short, the common sense(s) of the society of which they are part and product. The commitments bound up in them—like the Hobbesian notion of the state of nature—shape ideas, beliefs, and knowledges by delimiting the possible and denominating the unthinkable.

What this points up is the imperative of listening to voices whose own common sense(s) are radically different from our own. This performs two vital functions: it is a first tentative step toward the release of these marginalized voices from the obscurity imposed by the hegemonologue; and, it aids in exposing the indeterminacy of some of our own most fundamental "truths." Revelations of this sort, by denaturalizing hegemonic orders and ideas, aid in highlighting the vital contribution of the travelogues to the Enlightenment— much more than merely confirming key Enlightenment ideas, these accounts helped to constitute them by furnishing the negative definitions necessary for their full articulation. In unsettling the accounts of the travelogues, then, we simultaneously destabilize the hierarchies generated by a host of discursively gendered and racialized dichotomies borne in the hegemonologue: order/anarchy, culture/nature, rational/irrational, civilized/savage, to cite but a few.

As we have seen, scholarly accounts of pre-Columbian warfare in the Americas turn out to be startlingly indeterminate. They also quite flatly contradict Lakota autoethnographies describing cosmology and traditional lifeways—including the practice of warfare.[23] Surely, though, the validity of an account of any aspect of the aboriginal condition of a given people must, to the extent possible, be judged also in light of the sociopolitical, cultural, and cosmological contexts of that people. This calls for a more broadly intertextual approach, admitting some of the very voices silenced by orthodox treatments. It follows, then, that the knowledges borne in oral literatures must be acknowledged as being every bit as valid—and every bit as indispensable—as those preserved in the conventionally privileged written form. Listening to autoethnographical voices yields not only an account of the aboriginal condition of the Lakota that is quite different from those put forth by the anthropological and historiographical orthodoxies but also, as we have seen, an alternative conception and practice of political order that is equally at odds with that to which the orthodoxy of International Relations holds.

Given the shared assumptions of the various academic orthodoxies briefly considered herein, it should be of considerable interest that the widely accepted accounts characterizing the aboriginal condition of Indigenous peoples as mired in interminable warfare are not, as might have been imagined, founded on unambiguous evidence unmediated by subjective interpretation. Equally noteworthy is the dearth of contact between scholars working in Anthropology and History and those who make their disciplinary homes in International Relations. And yet, we find the orthodoxies of these relative solitudes mutually invested in ontological commitments that both privilege the state as the sole locus of political order and render the aboriginal condition of Indigenous peoples as anarchic. This is revealing of the politics of academic disciplinarity insofar as it highlights how, in important ways, these fields have never truly been separated. More programmatically, it points up not only the profound ethnocentrism of scholarship situated in the orthodox traditions but also the importance of confronting their sites of origin, the travelogues of Europe's Age of Discovery, as foundational texts of the social sciences. To read contemporary Realist-inspired International Relations theory without also reading Hobbes is to miss much in the way of the dubious foundations upon which the former has been constructed. Likewise, when we read the philosophers of the Enlightenment without also reading the accounts of missionaries, conquistadors, and colonial administrators, we risk missing their centrality to the canons. Even so, their voices can be heard echoing through the legacy of social contractarian thought as well as in contemporary orthodox social theory, the colonial purposes they serve(d) all the while obscured but very much intact.

In January 2000, the Grass Roots *Oyate* began what would become a lengthy though peaceful occupation of the offices of the tribal government on the Oglalas' Pine Ridge Reservation in South Dakota. Among the group's stated objectives was the abolishment of the tribal government imposed under the Indian Reorganization Act of 1934 and a return to traditional forms of political organization.[24] Their aspirations, then, were to (re)implement precisely that which orthodox social theory implicitly—and sometimes, as we have seen, explicitly—casts as implausible. The result is that Indigenous peoples are denied the possibility of a politics and are reduced instead to a political *issue*, itself confined to the domestic realm of the settler states in which they are situated. Such are the workings of advanced colonialism.

The immediate implications of this for people living on the reservation recall the instrumental use of Chagnon's writings by resource companies in Brazil. On Pine Ridge, as on many other reservations, IRA tribal governments have often been implicated in serious mismanagement in areas such as the administration of social programs and stewardship over local mineral, grazing, and dumping rights.[25] The extent to which tribal councils are free from the leading and/or limiting influences of the settler state is also in question since, as Biolsi points out, the IRA constitutions had written into them "provisions for review or approval by the Secretary of the Interior" of actions undertaken by the councils (Biolsi 1985: 657). In the first half of the 1970s, American Indian Movement (AIM) activists joined traditionalist efforts to unseat the Oglalas' IRA council headed by Richard Wilson and replace it with a reconstituted traditional form of political organization. These developments were met with the arrival of U.S. Marshals on the reservation and extensive interference with the impeachment campaign; a sustained wave of political violence, which left scores of traditionalists and AIM activists dead, continued for several more years.[26] In combination with the material deprivations that so regularly attend reservation life, these are the very real consequences of the denial of traditional Indigenous political possibilities.

The idea of the savage in the state of nature also fulfills a vital rhetorical function in support of the contemporary settler state itself. Juxtaposed against the self-ascribed virtues of Euro-American society, it justifies past conquests as well as subsequent and ongoing assimilative practices, even to the extent of making them seem morally imperative. Simultaneously, Indigenous peoples—or at least their aboriginal lifeways—are once more conflated with the natural challenges once offered up by the "untamed" terrain on the American frontier. Like the pop-culture accounts of Indigenous peoples, this phenomenon relegates all aspects of aboriginal lifeways—save, perhaps, for such markers and cultural accoutrements as have been appropriated into the semiotic performances of the settler state's own constructed identity—irretrievably to the distant past. This enables even those who might lament the (noble) savage's loss of natural freedom to accept it nevertheless as the inevitable result of "progress" and the steady march of "civilization"; in its contemporary manifestation, it renders the politics of the traditionalists as bewilderingly idealistic, even self-deluding. And all of these effects aid in the ideational production of Euro-American society and the modern Western Self. Even as it denies the possibility of a non-anarchical aboriginal condition, then, the Hobbesian impulse is essential as description of the bare life in opposition to which the virtues of the dominating society can be articulated. In this sense, we see again that the advanced colonial subjugation of Indigenous peoples is one with Euro-American self-knowledge(s).[27]

The incommensurability between the ethnographies produced by the academic orthodoxies and the autoethnographic accounts by Lakota people themselves is also revealing of the extent to which the former are racialized. The conceptual indebtedness of orthodox anthropological and historiographical treatments of Indigenous peoples to the travelogues inexorably involves them not only in the material aspects of colonialism/advanced colonialism but with the rhetorical constructs of the colonial encounter as well. As the European empires expanded into the rest of the world, a dialectical relationship took hold between racial ideologies and the exigencies of material exploitation, each impelling the other (Loomba 1998: 113). This, in turn, was elemental in defining an emergent knowledge system that endorsed the discourse of savagery and the attendant idea of an anarchic state of nature. Constructed in terms consistent with the cultural logic of the Age of Discovery, these ideas fed back into it, reconfirming themselves. It is in their adherence to vital aspects of these same ontological commitments that the orthodox anthropological and historiographical literatures are most profoundly racialized. And, notwithstanding that it might make no explicit reference to race, orthodox international theory is exposed as being similarly and unavoidably racialized for having built upon this same ontological terrain—a terrain defined by commitments reciprocally constituted by and constitutive of racial ideologies.

To the extent that orthodox theoretical approaches to International Relations exclude aboriginal knowledges and lifeways in deference to the familiar Hobbesian impulse, they are inseparable from the more comprehensive processes of invalidation by which the colonial subjugation of Indigenous people(s) is sustained. Though not directly culpable as purposeful agents, scholars working in this tradition, like their counterparts in the anthropological and historiographical orthodoxies, are nonetheless implicated in the ongoing project of advanced colonialism. It is in reproducing the hegemonic knowledges that invalidate many non-Western worldviews and lifeways that scholars working in these traditions exert a tyranny over Indigenous peoples. Articulated through research, writing, and (especially) teaching, the collective discursive power of orthodox scholars to define what is real, what is possible, and whose voices count can have considerable reach. The consequent denial of voice obscures the indeterminacy of dominant truth claims that, in turn, foreclose transformative possibilities and reconfirm the presumed naturalness of the hegemonic structures and ideas that enable the ongoing advanced colonial domination of Indigenous peoples. Here the inattention of the International Relations orthodoxy is as significant as the attentions of those of Anthropology and History—as Fabian reminds us, "writing need not have the Other as its subject matter in order to oppress the Other" (Fabian 1990: 767–68). Furthermore, if, as has been argued in this chapter, the commitments by which the denial of aboriginal knowledges might be justified do not stand up to critical scrutiny, we are left with the unsatisfactory circumstance that these selfsame commitments, by orienting the interpretation of ambiguous evidence, are themselves the sources of whatever putative proof can be invoked to support them. By extension, the invisibility of Indigenous peoples from the perspective of adherents to the orthodoxy of International Relations is in some measure reproduced by the failure of these same scholars to see them. Though the particulars of their cosmological commitments may not be generalizeable to other Indigenous peoples, the case of the Lakota traditionalists alerts us to the imperative of engaging our Others in conversation and, not least, to the dangers of allowing our own philosophical commitments and inclinations to foreclose a priori the very possibility of such engagements. It also calls upon us to recognize that International Relations theory is a powerful social force in its own right and is therefore susceptible of becoming an instrument of domination.

Notes

1. I make this point with considerable apprehension. While I think it important to note that the Indigenous peoples of the Americas have achieved a degree of standing in the international system, I do not wish to suggest that this ought to be the standard upon which the appropriateness of their inclusion in International Relations be judged. Similarly, I would be leery of any attempt to elevate Indigenous peoples in the popular imagination by way of reference to particular characteristics of social or political organization presumed as analogous or nascent forms of those of the dominating society. To do so is to fall into a form of evolutionist conjectural historicizing and to implicitly privilege Euro-American forms of social and political organization by making appeal to them as evidence of an "advanced" society.

2. See Winch (1990: 15).

3. A point of clarification is in order here: none of this is to say that orthodox commitments rule out the possibility of different forms of *organization*, only that non-state forms are not accepted as sources of *order*. For Hobbes, "composite

bodies politic had not really left the 'state of war' behind, they were only quasistates" (Forsyth 1979: 205).

4. Orthodox accounts of the aboriginal conditions of Indigenous peoples are in many ways much more expressive of the dominating society. As Edward Said has argued, "Orientalism is—and does not simply represent—a considerable dimension of modern political-intellectual culture, and as such has less to do with the Orient than it does with 'our' world" (Said 1979: 12). In a similar vein and with specific reference to ethnography, Timothy Jenkins notes the related reductive and productive functions of conceptual dichotomies: "The use of the paired terms modern/backward says more about the world of the enquirer than that of the peasant. In the stereotype, peasants are survivals of the pre-modern, embodying its qualities which are defined against our own" (Jenkins 1994: 450).

5. This bespeaks a more subtle and less instrumental working of colonial discourse than that suggested by Said's sense that imperialism and colonialism are "supported and perhaps even impelled by impressive ideological formations that include notions that certain territories and people *require* and beseech domination . . ." (Said 1993: 9; emphasis in original). This might tend too much toward what Homi Bhabha has called "the transparent linear equivalence of event and idea" (Bhabha 1990: 292). That is, it seems to imply too unitary a connection between colonial discourses and actual colonial practices. See also Loomba (1998: 232).

6. See, e.g., Holsti (1989) and Walt (1998).

7. See Suganami (1990: 21–23); see also True (1996: 229).

8. The choice of competition over cooperation, then, is revealed as being just that: a choice. And it is, as Johan Galtung has observed in a different context, an unduly pessimistic one at that. See Galtung (1969: 180–81).

9. Elsewhere, Gilpin appeals directly to the immutability of human nature as a constant underwriting cycles of hegemony and hegemonic decline (Gilpin 1989: 37).

10. See Keegan (1996).

11. As Mary Pratt notes, such renderings have been an important element in the legitimization of European conquest of Indigenous peoples (Pratt 1992: 186). And to this we might add that legitimization of this sort is as well conferred retrospectively as it was in the event.

12. See also his seminal work, *Y,anomamö, the Fierce People* (Chagnon 1968).

13. Chagnon's choice of the word "fierce" to describe the Yanomami (see chapter 4) has thus become a contemporary functional equivalent of the label "cannibal," used to such great effect as a normative inscription upon those Others who have stood in the way of colonial aspirations since Columbus's first voyage to the Americas in 1492 (See Motohashi 1999). Similarly, Loomba notes that "Spanish colonists increasingly applied the term 'cannibal' and attributed the practice of cannibalism to those natives within the Caribbean and Mexico who were *resistant* to colonial rule, and among whom no cannibalism had in fact been witnessed" (Loomba 1998: 58–59; emphasis in original).

14. See Brantlinger (2003).

15. In Holm's view, militarization of Indigenous peoples was "a method of assimilation or subjugation, depending on the viewpoint, equal to, or perhaps more effective than, that of outright military conquest, conversion to Christianity or economic dependency" (Holm 1997: 462).

16. That Ewers does not seem to have felt compelled to propose an answer to the question of whence these animosities originally sprang is, once again, suggestive of a prior assumption of a Hobbesian state of nature.

17. For perhaps the most comprehensive inquiry into early Euro-American accounts, see Berkhofer (1978).

18. See Sundstrom (1997). Winter counts are the basis of the traditional oral historiographical records of the Northern Plains peoples wherein each year is identified by way of association with some notable event.

19. Bamforth cites a count of at least 486, noting that perhaps 50 additional skeletons remain in place (Bamforth 1994: 106). P. Willey and Thomas Emerson offer a different explanation for the imprecision of the count: "Before the remains could be excavated by the USD Archaeology Laboratory, the remains of nearly 50 individuals were looted from the bank" (Willey and Emerson 1993: 265).

20. Noting that an incomplete second ditch had failed to enclose the expanded village before the massacre, Willey and Emerson speculate that the inner ditch fell into disuse because the village was not under constant threat of attack (Willey and Emerson 1993: 230–31). Such a view is consistent with speculation linking the massacre at Crow Creek to intermittent food shortages. It also reinforces the position that warfare was not endemic.

21. Though he is most accurately situated in the orthodoxy, Frank Secoy also notes that fortification increased in the post-contact period, observing that, "the art of village fortification, long in existence, had been developed to high efficiency in defense against both the gun-equipped northeastern peoples and the horse-riding southwestern ones" (Secoy 1966: 72). Note that fortification is found to have increased in response to adversaries equipped with horses and/or guns, both of which were introduced by Europeans.

22. As noted earlier, Ferguson makes no claim to the effect that warfare was absent from the aboriginal condition, holding only that the arrival of Europeans incited and intensified warfare amongst Indigenous peoples.

23. Lakota autoethnographies bear accounts at least as plausible as those of the academic orthodoxies. They are consistent even with the evidence of catastrophic conflict in the context of severe food shortages inasmuch as famine would almost certainly have bespoken a loss of cosmological balance. See chapter 4 of this book.

24. Personal communication, February 20, 2000.

25. See Churchill and LaDuke (1992); see also White Face (1998).

26. See Robbins (1992: 103–04).

27. The resultant identity–knowledge complex has proved remarkably resistant to critical reevaluation. Recall, for example, the hostile reactions to the Smithsonian Institution's "West as America" exhibit noted in chapter 5.

CHAPTER 7

Emancipatory Violences

Violence is a tricky thing: we see it when we know it, but we do not always know it when we see it. As Johan Galtung has famously argued, it need not be the product of malicious intent; it may as readily be structural in the sense that it is harm that is visited upon others in consequence of the political choices of the comparatively privileged and powerful but is not necessarily an objective of those choices in itself (Galtung 1969). We have seen how violences of this sort play out structurally in popular culture forms and in orthodox social theory notwithstanding that those realms might seem, on first gloss, to be quite remote from the advanced colonial structures of domination of which they are ultimately found to be an integral part. And even though there is no conscious instrumentality at work here, the outcomes are no less pernicious and their founding no more politically neutral than where more conspicuously malign violences are concerned: witness, for example, the parallels between Stevens's Sioux War Panorama and the Royal Ontario Museum's "Ontario Archaeology Gallery"; between the travelogues and some contemporary ethnographies of the Yanomami. Pathologies of this sort must be known before they are easily seen by those privileged enough not to suffer them. My aim in this chapter is to draw attention to some even more elusive violences than these in the hope that in coming to know them better we might also be better equipped to see them.

I have said that all international theory is cosmologically inflected in ways that make it complicit in ongoing processes and practices of advanced colonial domination in the Americas and elsewhere. To this point, however, I have engaged only the theoretical orthodoxy. But emancipatory approaches are also implicated in the violences of advanced colonialism; they too effect erasures and violently speak the hegemonologue over and against the voices of Indigenous people(s). It is for this reason that I have seen fit to argue that, in spite of its appearance of relative conceptual heterogeneity, International Relations is profoundly monological, speaking the voice of that knowing Western subject whose universalist pretensions generalize and naturalize the concepts, categories, and commitments of the dominating society. More broadly, aspects of all of these are also spoken in unison with other voices of the hegemonologue: as heard through the varied representational technologies of popular culture, for example. To be sure, emancipatory approaches unsettle and dispute a range of politicized claims borne in the uninterrogated assumptions and "common senses" of the dominating society. Others still, however, not only are missed by their various critiques but also are actually carried forward and reproduced by them. It is to these that I turn now.

Before proceeding, though, a few words are in order regarding what I am here identifying as "emancipatory theory." Some readers might be surprised to find that I lump liberal-inspired approaches together with the more thoroughgoing critiques of the sociopolitical status quo developed by Marxists, feminists, and others. This move is not at all to downplay the important ways in which the points across so broad a theoretical spectrum are differentiated. Rather, it is to highlight their one essential shared trait: they are all inclined in some way toward the emancipation of the socially and/or politically marginalized, even if their accounts of what that might mean are sometimes quite profoundly at odds. Liberals might promise emancipation through the very structures and processes identified as the sources of marginalization by some of the deeper critiques but, lest we give in to the vulgarity of cynical caricatures, theirs is a sincere undertaking nonetheless. In this sense, it is noteworthy that all emancipatory approaches—liberal ones included—turn vitally on some notion of "the good life" even if it might not always be clearly specified.[1] At the same time, I accept Wight's persuasive argument that orthodox international theory has naught to do with the pursuit of "the good life," its pessimistic view of the world having fashioned a "realm of recurrence and repetition" wherein survival is the closest thing to a "good" (Wight 1966: 29).[2]

This fundamental difference of orientation marks out the orthodoxy as distinct from all other theoretical approaches, necessitating its treatment in isolation. For this reason, it might seem as though the line of critique I have been developing is now somewhat disproportionate in application. After all, I have dedicated a chapter to uncovering the advanced colonial complicities of Realist-inspired international theory, and now propose to devote only equal measure of attention to all other approaches combined. True enough, the Realist tradition continues to cast a large shadow over disciplinary International Relations. But I wish neither to privilege the orthodoxy nor to suggest that no other theoretical approach is of sufficient import to warrant the same degree of attention. Conversely, there is no ulterior motive at work here to shield emancipatory theories from a full measure of critical scrutiny. On the contrary, the critique that follows touches all emancipatory approaches to international theory, even if to varying degrees. To the extent that some of these are also implicated in more particular erasures of the sort effected by orthodox social theory, I am able to draw on arguments already proffered without retreading the same ground. Similarly, their mutual implication in what I am calling "emancipatory violences" permits an economy of critique of a range of theories through a focus on their points of intersection, rather than engaging separately with each.

The Tyranny of Emancipatory Social Theory

For those whose theoretical commitments wed them to a vision of a more equitable world, the idea that the advanced colonial violences of the hegemonologue work even through emancipatory theory is a bitter pill to swallow. But it is not entirely novel either. Feminists went through something similar in the early-1980s as the disappointments of feminist-inspired development initiatives in the non-Western world forced a critical rethinking of the universal category of "women."[3] In particular, the experiences of White, middle-class Western women were found to have been inappropriately generalized in ways that defied alterity and disabled the fashioning of a political praxis consonant with local contexts. The unified voice sought by some feminists thus turned out to speak from a position of conspicuous privilege with the result that it subverted the emancipatory potential of its own founding by asserting an idealized Self over and against Other women.[4] Consequently, what had been conceived as a unified oppositional subjectivity was progressively exposed as hegemonic in its own right by women in the developing world—and, increasingly, within the West itself—who objected that it did not include or reflect their experiences.

Even more problematic, such distinctions as were made tended to map neatly with global sites of margin and privilege in ways that seemed to underwrite the West's hegemonic knowledges about its Others. Indeed, as Chandra Talpade Mohanty argues, the prevailing constructions of the singular "Third World Woman" readily betrayed underlying colonial discourses through which the Western Self and non-Western Other have been codetermined (Mohanty 1984: 334). As she describes it,

> a homogeneous notion of the oppression of women as a group is assumed, which, in turn, produces the image of an "average third world woman." This average third world woman leads an essentially truncated life based

on her feminine gender (read: sexually constrained) and being "third world" (read: ignorant, poor, uneducated, tradition-bound, domestic, family-oriented, victimized, etc.). This, I suggest, is in contrast to the (implicit) self-representation of Western women as educated, modern, as having control over their own bodies and sexualities, and the freedom to make their own decisions. (Mohanty 1984: 337)

An emancipatory agenda does little to temper Self/Other constructions that are so uncomfortably reminiscent of colonial-era discursive representations. Aihwa Ong puts it thus: "By portraying women in non-Western societies as identical and interchangeable, and more exploited than women in the dominant capitalist societies, liberal and socialist feminists alike encode a belief in their own cultural superiority" (Ong 1988: 85). All of this suggests that Spivak's warning about the voicelessness of the subaltern is salient also as a critique of global gender categories, in both the essentialization of identity and the universalization of a particular (privileged) experience.

Emancipatory international theory is no less implicated in processes of erasure. Like orthodox theory, it typically makes no mention of either Indigenous peoples or advanced colonialism. Even postcolonialism has done little to address these omissions. Said, as we have seen, consigns direct colonialism largely to the past. And recent postcolonial scholarship on the whole seems to bespeak a similar assumption, tending to focus more upon the enduring effects of colonialism than on the effects of enduring colonialism.[5] It is also noteworthy that emancipatory theory more generally, and emancipatory international theory in particular, has failed to distinguish the settler state as a distinct type—a lapse that obscures the fact of ongoing advanced colonial domination. Just as the orthodox denial of non-state possibilities for political order effects erasures, so too emancipatory approaches are implicated in a range of advanced colonial violences. Most noticeably, they (re)produce the exclusion of Indigenous peoples from International Relations, sometimes more or less passively (through simple inattention) and sometimes actively (through constructions of their own making).

Foremost among the ideational constructions effecting erasure are the grand narratives of Enlightenment tradition. These have, in various incarnations, been central both to articulations of "the good life" offered up by emancipatory theories and to the varied means prescribed to attain it. The universalized concepts at the heart of these narratives are familiar: individualism, secularism, class, gender. But though they might be well enough known to us in their particulars, their violent complicity in oppressions is seldom adequately acknowledged, even as they are enlisted in aid of ostensibly emancipatory designs. In the very pretension of universality is an implicit denial of the contested, and oft times violent, forging of concepts such as these. Bhabha puts it thus:

> I would like to suggest that these concepts of a modern political and social lexicon have a more complex history. If they are immediately, and accurately, recognisable as belonging to the European Enlightenment, we must also consider them in relation to the colonial and imperial enterprise *which was an integral part of that same Enlightenment*. For example, if liberalism in the West was rendered profoundly ambivalent in its avowedly egalitarian project, when confronted with class and gender difference, then, in the colonial world, the famous virtue of liberal "tolerance" could not easily extend to the demand for freedom and independence when articulated by native subjects of racial and cultural difference. Instead of independence they were offered the "civilising mission"; instead of power, they were proffered paternalism.   (Bhabha 1996: 208–09; emphasis in original)

In seeking to build an emancipatory project upon concepts such as these, then, their historical implication in structures and processes of oppressive domination is obscured from view. Likewise, the inherent violence of any presumption that they are ubiquitous seems to go unnoticed—the emancipatory potential of each of these concepts and their attendant discourses is too often conditional upon a sweeping aside of that which would deny their universality. Emancipation is thus founded upon tyrannies of its own.

Liberal international theory is quite clearly illustrative in this regard. Whether in its prescriptions for world peace or its conceptions of "the good life," it follows Realist-inspired international theory in flatly contradicting autoethnographic accounts of Lakota cosmology and traditional lifeways. Liberal social theory in general privileges the private sphere of individualized rights and denigrates more communally oriented arrangements of the sort expressed through the *tiyospaye* and in the buffalo virtues. It is also founded upon a decidedly Western notion of progress that despises the natural world. This not only stands in stark contrast with the commitments bound up in a worldview consonant with the sacred hoop, but has a long history of direct entanglement with the business of colonialism as well. If, as Locke pronounced, "in the beginning, all the world was America" (Locke 1924: 140), then European superiority seemed in yet another sense to have been confirmed, and this could confer legitimacy upon the taking of Indigenous peoples' lands. Locke proposed that the limit on the extent of land that could be counted as one's own was equal to whatever measure one was able to "improve" and cultivate (Locke 1924: 132). European colonists could thus feel justified in their seizure of lands in the Americas since the original inhabitants, having failed to make "improvements" to those lands by means of their labor, were thereby imagined to have forfeited any claim to a right of property over them. And if the Americas had theretofore been left to lie fallow, their colonization and cultivation by Europeans could only be to the benefit of humanity in general.[6] It should therefore be of more than just passing interest that, as we have seen, aggregate "Indian" stereotypes have drawn principally on the itinerant Plains peoples and not the many sedentary agrarian societies of the pre-Columbian Americas—as Gene Weltfish has observed, "[t]he universality of the Plains Indian nomadic stereotype as a function of the natural order of things gives active support to this [Lockean] ideology" (Weltfish 1971: 222).

Contemporary liberal international theory continues very much in this tradition. In particular, it remains committed to individualized conceptions of liberty expressed most conspicuously in property rights. Partly as a consequence of growing dissatisfaction with Neorealism, liberal theory has enjoyed a renewed popularity among International Relations scholars since the end of the Cold War. Rejecting the cold stasis of the Realist power-politics calculus in which conflict is the unswerving first feature of human social interaction, the new liberals take a view of politics as, in Robert O. Keohane's words, "open-ended and potentially progressive, rather than bleakly cyclical" (Keohane 1989: 11). It is thus that a space is created in which to imagine an emancipatory agenda operationalized through cooperative interaction. Liberalism's accounts of processes and phenomena such as international cooperation or interdependence turn on a presumed universal pursuit of "the good life" that is ultimately defined in terms of material gratifications. Importantly, however, the idea of progress works through this such that it is driven by material wants that can never truly be satisfied. Accordingly, subduing the natural world and harnessing it to productive purposes through industry is a *sine qua non* for the enjoyment of "the good life." And the ideational terrain upon which this relationship is sustained is marked out in advance by those elements of the hegemonologue—with their origins in the Judeo-Christian Creation story and Plato's privileging of the rational over the experiential— that, as I have argued, cast nature as the prototypical Other.

At the same time, liberal ruminations on the transcendence of war work indispensably through the state, thereby repeating some of the violences of erasure that I have attributed to the theoretical orthodoxy of International Relations. Following at least the broad strokes of Immanuel Kant's *Perpetual*

Peace, democratic peace theory suggests that war would be an uncommon enterprise in a world populated only by liberal democratic states. Building on the thesis that liberal democracies are inherently less prone to war (owing variously to inhibiting structural characteristics and/or normative commitments), contemporary exponents of this thesis have increasingly put the accent on democratic institutions through which states cooperate as well as the interdependence that results from trade and economic integration.[7] Clearly, traditional Lakota forms of sociopolitical organization are not at all well accommodated by this. Once more, the state is, in the same instant, naturalized and valorized—this time as the form of political community through which a lasting peace might be realized. Though certainly much less direct in this sense than the constructions of savagery in which Realist-inspired theory is implicated, democratic peace theory enacts its own erasures nonetheless. Moreover, its emancipatory promises turn out to be conditional: having just offered that, "liberalism's ends are life and prosperity, and its means are liberty and toleration," John M. Owen cautions that this requires that people be enlightened and aware of their interests and that they "live under enlightened political institutions which allow their true interests to shape politics" (Owen 1994: 94). Given the focus here on liberal democracy articulated through the state, one need scarcely wonder at what might be counted as "enlightened political institutions." And this assimilative impulse is repeated in liberalism's material account of prosperity. Emancipation, it seems, is here again contingent upon a fundamental commitment to rendering sameness. In extreme formulations, this tends strongly toward celebrating global cultural homogenization and the subordination of diverse value systems to the profligate consumption of ever more goods and services.[8]

Again, the voice of the hegemonologue in this account of "the good life" is audible in the characteristic Western cosmological siting of humans apart from and above nature. It is therefore quite telling that this inclination is discernible also in other emancipatory theories. And the homogenizing impulse borne in emancipatory designs turns out to be a more general phenomenon also. As Rice observes:

> The hatred of the physical world finding expression in the most seminal western thinkers from Plato through Paul to Augustine to Calvin reaches into the philosophies of Freud and Marx. Freud thought of nature as a tooth and claw battle of beasts, and of human nature in a state of continual war with an inner beast, requiring Calvinistic vigilance. Marx had little sense of existence apart from abstract economic and social arrangement,

a view from the air. . . . Marxism is not the first "great religion" to be "ethnocidal by design." Its intention to level national differences emulates the Christian effort to erase religious ones.   (Rice 1991: 26)

Marxist-inspired accounts of "the good life" are underwritten by this view of humanity's relationship to nature: in common with liberalism, a Western notion of progress merges here with subordination of the physical world to the satisfaction of material wants. As Rice points out, "Marxism replaces the dualism of Christian society [i.e., heaven and earth] with its own dualism of matter and spirit, by valorizing matter" (Rice 1991: 24). And as with liberalism, this implicates Marxists in the ideational structures of historic colonialism to which Indigenous peoples have been subjected.[9] "In certain respects," according to Loomba, " 'progress' was understood in similar ways by capitalists as well as socialists—for both it included a high level of industrialization, the mastery of 'man' over 'nature,' the modern European view of science and technology" (Loomba 1998: 21). All of this, of course, is also quite contrary to traditional Lakota accounts of "the good life," understood as the imperative of maintaining balance.

The rigid teleology of much Marxist-inspired theory is especially prob-lematic to the extent that, in the same way as liberal notions of progress, it ends up bespeaking the inevitability (and, again, the desirability) of the dis-appearance of Lakota traditionalism. Friedrich Engels's notion of "primitive communism," for instance, is elaborated in the past tense as a first form of human society.[10] But it turns out that this idea, like the orthodox social theorists' accounts of life without the Leviathan, rests upon bad ethnography that, in turn, implicates it in the very social evolutionism that has underwrit-ten so many colonial violences.[11] Implying all that we might expect the word "primitive" to connote, and as befits its founding in a progressive discourse, it is a condition ultimately to be transcended. This is significant to the extent that some important aspects of traditional Lakota lifeways are crucial to Engels's description of "primitive communism": most notably, the absence of both social hierarchy and a productive surplus. That these are taken as the defining features of a first form of human social organization and produc-tive endeavor is quite telling given that they are constructed by Engels as something inevitably (to be) transcended. Progress, then, brings in tandem the unsustainable immoderation of surplus production. And in its discursive framing, this has the effect of valorizing excess while sustainable lifeways are, somewhat paradoxically, cast as unsustainable in an historical sense.

According to Clastres, the problem here is fundamentally that what have been inscribed as "primitive" societies cannot be what they were/are because Marxism insists on their being what it needs them to be: in a word, "precapitalist" (Clastres 1994: 136–37). In his retrospective on the violences of the European "conquest of the planet" after 1492, Samir Amin argues that prior to the colonial encounter, "gestating" capitalisms existed in the non-European world and that this reflected "a general law of evolution of human societies" (Amin 1992: 12). And lest there be any uncertainty about the extent of this claim, Amin holds that "[f]ar from having introduced capitalism to the peripheries of global capitalism, the Western expansion sometimes delayed its ripening and always deformed its development so as to create an impasse" (Amin 1992: 14). But the idea that nascent forms of capitalism were "gestating" throughout the world is a profoundly teleological form of homogenization that denies Lakota cosmology. Moreover, there is a clear sense in which this might serve as an apology for Europe's conquests: if capitalism was everywhere inevitable, then we might imagine that colonialism only hastened an outcome that was, in any event, preordained. Capitalism is a system of exploitation whose very logic demands expansion, meaning that sooner or later those "other capitalisms" would have had to insert themselves into the capitalist world system; the more advanced capitalisms of Europe would then have exploited them on these universal terms as surely as they did through direct colonial control. We might be forgiven for wondering, then, just what informs Amin's complaint, since the outcome is the same either way. Perhaps it is enough to lament that the rest of the world was not permitted to be the architect of its own exploitation. Regardless, Amin leaves no room to doubt its eventual insertion into a capitalist world system dominated by the advanced capitalism of Europe. And it is thus that comfort could be given to colonialism's apologists, insofar as direct conquest would seem only to have hastened the inevitable.

Even notions like alienation leave Lakota traditionalism either problematically inscribed or, alternatively, excluded. Noting that alienation is a central problematic taken up by Christianity as well as by Marxism, Deloria argues that it reflects a cosmological predisposition not generalizable to Indigenous North American societies:

> Indians . . . are notably devoid of concern for alienation as a cosmic ingredient of human life, a question to be answered or a problem to be confronted. This is not to say that Indians do not *feel* some degree of alienation. Rather, they do not make it a central concern of their ceremonial life, they do not feature it prominently in their cosmic mythology, and they do not see it as an essential part of institutional existence which colors their approach to other aspects of life. Alienation, therefore, is an essential

element of Western cosmology, either in the metaphysical sense or in the epistemological dimension; it is a minor phenomenon of short duration in the larger context of cosmic balance for American Indians. (Deloria 1983: 114–15; emphasis in original)

This view is shared by Frank Black Elk, who offers that the traditional Lakota commitment to the interrelatedness of all in Creation—expressed as *mitakuye oyasin*—destabilizes Marxist conceptions of alienation:

> We, as a people (within the traditional culture view, at any rate) view ourselves only in direct (natural) relation to everything else at all times. Thus, we *cannot* feel the sort of distance indicated in the notion of alienation, either between each other as people, or between ourselves and any aspect of the universe. Alienation is an impossibility within traditional Lakota culture; we are prevented, by the way we view reality, from taking those steps which would, sooner or later, produce the condition of alienation. (Black Elk 1983: 152–53; emphasis in original)

And this is suggestive of how it might be argued that traditional Lakota cosmology is actually truer to the idea of dialectical knowledge than Marxist-inspired theory: *mitakuye oyasin* resists the unequal oppositional rendering of ontologized binaries that Marxism ultimately upholds in its denigration of nature.[12]

What this signals is that the hegemonologue speaks here too, subverting the possibility for a truly relational dialectics in deference to the easy temptations of oppositionalism—the binaries of Western metaphysics thus occlude their own radical potential, settling on the very categories that have enabled the violences of colonialism/advanced colonialism. Core/periphery, for example, is yet another oppositional binary, reproducing the separation of subject and object, Self and Other. Of course, it is useful heuristically, but when ontologized it can lead to pathologies reminiscent of those suffered by Spivak's voiceless subaltern. All of this, however, might have less to do with emancipatory discourses, per se, than with their apparent inability to recognize that they bear the voice of the hegemonologue, much less effect a silencing of it. (As Loomba argues, "[w]e need to move away from global narratives not because they necessarily *always* swallow up complexity, but because they historically have done so, and once we have focused on . . . submerged stories and perspectives, the entire structure appears transformed" (Loomba 1998: 249; emphasis in original). The inherent violences of Marxist-inspired theory, then, follow not so much from its emancipatory designs as from the combination of its (Western) cosmological boundedness and the pretension, born of universalism, to dictate its terms of emancipation without sufficient regard to specificity.

We see this quite clearly in feminist theory as well, though, as noted earlier, feminists more generally are not altogether unaware of the problem. Despite their best-conceived emancipatory designs, however, the hegemonologue can be found working through feminist theories such that they too may violently speak into being Western concepts, categories, and philosophical commitments; they too can be found to effect erasures. In particular, there is a notable tendency, manifesting in various ways, to denigrate the rendering of difference. For the more liberally inclined, the articulation of gender as difference—especially when mapped together with more or less discrete social roles and identities—is irredeemably oppressive, quite unambiguously marking out the necessary terrain upon which an emancipatory project must be founded. That is to say, the terms of emancipation are posed as the imperative ascription of an essential sameness, privileged over any identifiable elements of difference, such that the assignment of separate and specific roles to women and men respectively becomes unsustainable.

The disjuncture between this perspective and the idea of the sacred hoop could scarcely be more pronounced. The cosmological commitments expressed as *mitakuye oyasin* lend neither to the denigration of difference nor, certainly, to any suggestion that difference can or should be transcended. To be sure, different roles and characteristics are attributed to people according to their gender. But two important qualifications should accompany this observation. First, in the worldview of Lakota traditionalism, sex need not be determinate of gender. Contrary to the norms and conventions of the dominating society, gender ascriptions have been unproblematically bridged by persons who have felt a greater affinity with the social roles normally associated with their opposite sex. Indeed, there is even a conception of a third gender, personified in the *winkte*.[13] Unlike in the dominating society, social disciplining practices have not traditionally been brought to bear against those who transgress gender norms in this way.[14] On the contrary, *winkte* have occupied a valued social space and have normally been at least as esteemed as all other members of the *tiyospaye*: a bona fide third gender, every bit as essential to the social whole as the others.[15] According to Robert Fulton and Steven W. Anderson,

> gender was not necessarily determined by birth; it was something that could be acquired and, at times, changed according to the needs of the society. . . . Whereas the Western perspective of gender places "female" and "male" on a hierarchical continuum (at polar extremes with the "male"

dominant), preconquest aboriginal peoples, frequently egalitarian, typically understood gender as circular, (Fulton and Anderson 1992: 607)

Gender, from this perspective, is determined by neither sex nor sexuality.[16] Moreover, it is not a rigid category whose boundaries defy transgression, because masculine and feminine are inseparably as one rather than being held in radical opposition across linear space.

In keeping with the cosmology of the sacred hoop, there is no basis for disparagement of the difference signified by *winkte* (Bunge 1984: 43–44). And this extends to gender difference more generally. True enough, particular social roles typically belong to each gender, but, as the place (indeed, the possibility) of *winkte* demonstrates, people are not imprisoned by gender constructs, except to the extent that acculturation has implanted the attitudes of the dominating society.[17] Furthermore, as we have seen, the transcendent unity of a cosmos expressed as the sacred hoop valorizes alterity and resists the construction of hierarchy. We impose our own cosmological commitments, then, when we mistake social difference for inequality.[18] Consequently, to the extent that Euro-American feminisms imply that Lakota gender relations must manifest as oppressive hierarchical structures, they participate in the violences of erasure.

And yet, Lila Abu-Lughod tells us of feminist discourse in general, that "[i]t is becoming increasingly recognized that perhaps the very system of difference constitutes sexism and thus has to be understood and dismantled" (Abu-Lughod 1990: 25). She does, to be fair, precede this with an inferred caveat that can be read as a specification of a particular context in which this claim holds: "As Simone de Beauvoir pointed out so long ago, women have, *at least in the modern societies of the west,* been the other to men's self" (Abu-Lughod 1990: 25; emphasis added). We see something similar in Peterson's demystification of the violent mappings of binary oppositions:

In the Athenian polis, objectivist metaphysics (dualisms in Western philosophy), essentialized sexual identities (mutually exclusive male and female principles), and hierarchical state making (exploitative social relations) were mutually constituted. As one consequence, the public-private dichotomy legitimating the political order maps on to the culture-nature (subject-object) dichotomy of objectivist metaphysics, legitimizing domination practices (such as "objectification" of women, nature, "other") more generally. Implicit in the dichotomy of culture-nature (subject-object) is the intention of domination or control—fending off the unpredictability or instability of nature by imposing predictability and order through the power of classificatory systems and/or actual physical control. The construction of maleness underpins these dichotomies. Through the male-identified capacities for reasoning, abstracting, and formalizing, the man/masculine identity becomes the agent/subjectivity uniquely capable of transcending the realm of necessity—understood as nature, material and sensual embodiment, and concrete reproduction (femaleness/woman/feminine). (Peterson 1992: 36–37)

The value of these observations ought not be understated. But their rootedness in objectivist metaphysics traceable to the ancient Athenian polis bespeaks their specificity as well. To be sure, it is well that such a critique should be raised if for no other reason than the emancipatory potential inherent in laying bare the perverse matrices of meaning articulated through *différance*. At the same time, however, it is once again worth noting whose emancipation is at stake here.

A circular cosmology, as we have seen, defies binary opposition with the result that, where it is operant, the culture/nature dichotomy identified by Peterson cannot be sustained. And this, in turn, underscores that Lakota traditionalism's ascription of particular roles to women and men is not properly read as sexist or oppressive. The same can be said of many other Indigenous contexts.[19] This is not to say that power is not articulated through different gender roles, only that it does not conform neatly to the usual circuits anticipated by Western metaphysical predispositions. Once more, then, the intertext is revealed to be an intratext: an enclosed discursive space within which only the relatively privileged voices of the academy are spoken. As we have seen, this is a cosmologically bounded space wherein the possibilities inherent in traditional Lakota cosmological commitments can scarcely be imagined. The result is that in raising what is a critique of the gendered nature of Western modernity, feminist poststructuralism ends up effecting erasure through exclusion. Of course, an obvious objection here is that colonialism implanted the social arrangements and attendant ideational constructs of Europe throughout the world. But this once again reduces Europe's Others to passive skin, occluding the role of agency and resistance—a potential pathology, I have argued, of postmodernism and poststructuralism more generally. Thus, the most worthy effort to denaturalize hierarchically ordered binary constructions, it seems, may have the unintended effect of generalizing them.

Like postmodernism and poststructuralism writ large, feminist poststructuralism is indispensable as a critique of the foundations of the hegemonologue. If, however, this is allowed to substitute for conversation—that is, if equal room is not opened up for Indigenous readings of the hegemonologue—then it too has become complicit in a violence. As with other emancipatory approaches, the fundamental failing here is that conversational space has not been meaningfully broadened, with the result that de facto boundaries have been fixed upon the intertext of international relations.[20] Otherwise well-founded oppositional discourses thus devolve into essentially paternalistic pretensions to speak for the Other such that they can do no better than to dictate the terms of emancipation on behalf of marginal voices. The problem, of course, is that, as we have seen, speaking on behalf of Others turns out to mean speaking in place of them. Replicating the logics (and pathologies) of ethnographic (re)presentation, this reconfirms the Others' perpetual absence. The process is, by now, familiar: slotted into spaces marked out in advance by a cosmologically circumscribed range of Western concepts, categories, and (now) cognitive frameworks, Indigenous knowledges are remade in ways that divest them of their inherent emancipatory potential as radically oppositional discourses. Frustrating the fashioning of an ethics of responsibility such as I have suggested that it must be conceived, responsibility is here cast as *for* the Other (as ward) instead of *to* the Other (as co-conversationist). And, as I have argued, this is rightly counted among the violences of advanced colonialism inasmuch as it necessarily entails either the denial or the appropriation of Indigenous knowledges—a distinction without any difference since appropriation amounts to denial.[21]

I should add most emphatically that my aim in raising this line of critique is not the wholesale dismissal of this and other emancipatory projects. Rather, it is to highlight the boundedness of the terrain upon which they have been raised. The point, then, is not that feminist poststructuralism is somehow wrongheaded, but that it has thus far been taking place within cosmologically defined limits and that the failure to explicitly acknowledge these limits is problematic. Importantly, this problem is of a sort that feminists have identified elsewhere—as Jacqui True points out, for example, "[t]he habitual refusal of postmodernists to situate themselves as gendered theorists and grasp the transformative potential of—as well as their indebtedness to—'subjugated knowledges' such as feminism, traps them in the same gendered, hierarchical dichotomies that they claim to invert or transform" (True 1996: 242). We might say something similar of any theoretical approach that does not self-consciously situate itself cosmologically. Having exposed the same kind of problem as it pertains to gender, feminist poststructuralism is very well positioned to address the cosmological myopia of emancipatory theory more generally. To be clear, then, this is not a fundamentally flawed approach—unlike liberalisms or Marxisms whose ontological commitments contradict a range of Indigenous knowledges, it simply has not as yet been carried far enough. And where it does not extend to a consideration of cosmological diversity, it runs the risk of complicity in erasures that are similar in many respects to those it has so indispensably exposed.

Alchemy and the Perils of Liminality

Postcolonialism, of course, is not without problems of its own, beyond those briefly noted at the outset of this chapter. Indeed, I have reserved the most extended caveat for my chosen theoretical framework, offering something of a cautionary tale on the potential for advanced colonial violences to be mystified by a central conceptual commitment of recent postcolonial theory: hybridity. Recounting how the dominant Ladino group in the Guatemalan highlands has appropriated the idea of hybridity in an effort to delegitimize Mayan cultural activism, Charles Hale calls for an approach to theorizing that subordinates generalized theoretical concepts and categories to the specificities of the particular case:

> The "militant particularism" of this approach replaces grand theorizing with analysis arising out of negotiation and dialogue. It redirects attention to the pointed empirical questions about discourses of identity, which reach well beyond their allegedly "hybrid" or "essentialist" characters: who deploys them, from what specific location, with what effects? (Hale 1999: 313)

This case is revealing of the dangers inherent in all generalizations. Hale's proposed remedy, however, is doubly recommended here, simultaneously addressing another potential pathological effect of foregrounding hybridity: that in some instances it might conceal advanced colonial violences and hidden workings of the hegemonologue in some of the last places we might expect to find them. As a defining metaphor, hybridity implies a growing together, as in hybrid organisms. But as we shall see, not every hybridization is viable—sometimes the graft, or the host, or both wither and die for the effort.

Still, it remains one of the most valuable contributions of postcolonialism that it allows us to access and think about hybrid identities, acknowledging that colonizer and colonized alike are indelibly marked by the colonial experience. This, in turn, enables the important revelation that the colonizers can also be read as "camouflaged victims" (Nandy 1983: xvi) of colonialism. We are thus able to fashion a more nuanced account of colonialism/advanced colonialism wherein the specificities Hale bids us consider need not be banished from sight. Pratt's notion of the contact zone, for example, makes it possible to "foreground the interactive, improvisational dimensions of colonial encounters so easily ignored or suppressed by diffusionist accounts of conquest and domination" (Pratt 1992: 7). The emancipatory potential of this resides in its radical destabilization of the fixed identities that have, through both instrumental design and the more subtle workings of the hegemonologue, formed the core of the essentializing discourses according to which it has been possible to confine "different types" of people to particular temporal, spatial, and disciplinary contexts. Likewise, it has aided in recovering the space for an oppositional politics operationalized through practices such as mimicry. But as valuable as these insights are, we take our interest in hybridity too far, outweighing its contributions with a host of new pathologies, if we make the mistake of uncritically accepting all hybrid identities. To do so is to lose sight of the specificities of the politicized contexts in which hybridizations manifest and, therefore, to risk substituting a hegemonic agenda for an emancipatory one. In our empirical investigations, then, we must take care not to neglect the particular in deference to a more generalized notion of hybridity. Put another way, the idea of hybridity should not be conceived in a way that might validate any admixture.

A case in point, the phenomenon of "whiteshamanism" has become increasingly widespread in recent years as the ceremonies and semiotic markers of Indigenous cultural and spiritual traditions have come into vogue in a range of contexts and settings in the dominating society: from pop-culture feminism to a somewhat puerile fascination with the stereotypically derived regalia of Indigenous cultures exhibited by so-called hobbyists. Gathering at mock pow-wows garbed in feathers, beads, and buckskins, the hobbyists perform what they presume to be historically accurate reenactments of the traditional lifeways of peoples such as the Apache, Crow, and Lakota. Complete with self-styled "Indian" names, they role-play on weekends and holidays, assuming personas quite at odds with their otherwise typically middle-class lifestyles. For the most part, their motives seem benign enough, and many hobbyists profess a deep admiration for the people(s) they portray, coupled with a genuine desire to engage reverentially with decidedly non-Western knowledges and lifeways. As the organizer of a hobbyist event in Germany explained, "We are not copying them, or mocking them, but trying to really feel for this culture."[22] But however sincere the sentiment, hobbyism is not at all harmless. Though they may seem rather innocuous on the face of it, the hobbyists perform a function that is inextricably bound up with advanced colonial domination: the rendering of Indigenous people(s) as spectacle, as unreal.

As we have seen, severing Indigenous people(s) from their contemporary contexts, imagining them only in the ethnographic present and then only in terms influenced by aggregate stereotypes, (re)produces foundational knowledges of advanced colonialism. In creating a sense that the terms of their relationship with the dominating society were irreversibly forged on the battlefields of an earlier era, while simultaneously reinforcing the invisibility—or, at least, the unrecognizability—of real Indigenous people(s), such portrayals obscure their ongoing struggles. And in spite of whatever laudable sentiments might impel them, the hobbyists are deeply implicated in this process. In this regard, it is not insignificant that their portrayals are rooted in the (invented) past, typically identifying most closely with the mounted Plains warriors of the nineteenth century. Nowhere are there hobbyists who appear to be imagining themselves on a reservation at the dawn of the twenty first century; none seem interested in "trying to really feel" for the poverty, health problems, or, in some cases, the political violence that so often characterize the contemporary realities of the real traditionalists whose values they claim to share and whose lifeways they seem otherwise eager to experience.

Hobbyism, like the figural-discursive representations of popular culture, is troubling also because of what it tells us about the perspective on the relationship between Indigenous people(s) and the dominating society as it is cast from within the latter. Joel Monture, in describing a "pow-wow" he witnessed in Uckfield England, recalls how he found "people playing 'Indian' in much the same way minstrel shows once portrayed African-Americans for the amusement of themselves and others, with the accompanying stereotypes" (Monture: 1994: 114). Wondering why such conduct is not generally regarded as disrespectful or outrageous, Inés Hernández-Ávila asks us to "imagine people wanting to find out what it 'feels like' to take part in the Catholic ceremony of the Eucharist, or to wear a priest's garments, or the dress and hairstyle of Orthodox Jews, because it seems 'cool' " (Hernández-Ávila 1996: 343–44). These comparisons are in no way cavalier—it is not, after all, difficult to imagine that hobbyisms devoted to the cultures or spiritual commitments of many other peoples might ignite considerable outrage, not only among members of the group being imitated but in the dominating society more generally. The fact that the recreational affectation of Indigenous identities elicits no such generalized indignation bespeaks, on a very fundamental level, the consummate objectification of the peoples from whom they have been appropriated. This, in turn, is what makes possible the rendering of Indigenous people(s) as spectacle, admitting of hobbyism even as, in reciprocal fashion, hobbyism contributes to its reproduction. And inasmuch as this affirms a particular relationship between subject and object wherein Indigenous people(s) are denied independent ontological significance, it is inseparable from the broader processes that sustain advanced colonialism.

Notwithstanding that many followers of the New Age, Men's, and pop-culture feminist movements do not confine their dabbling in Indigenous cultures to weekend "pow-wows" and may genuinely seek to make Indigenous spiritual practices relevant to their own lives more generally, their conduct has effects arguably even more pernicious than that of the hobbyists. Like the hobbyists, they too have been accused of contributing to the objectification of Indigenous people(s) by perpetuating stereotypes. And like the hobbyists they are implicated in the broader processes of advanced colonialism to the extent that they implicitly—and often explicitly—assert that they have a right to appropriate Indigenous spiritual practices for their own uses. This typically is advanced in tandem with the supporting proposition that Indigenous spirituality cannot rightly be claimed as the cultural or intellectual property of any particular people. This view, of course, is wholly consistent with the deeply ingrained Western philosophical privileging of individual rights over communal rights—a disposition that is itself fundamentally inimical to many of the traditions being appropriated. Whitt compares such claims and their deeper individualist underpinnings to the doctrine of *terra nullius* by which the European colonial powers justified their conquest of the Americas: "The notion of property belonging to no one is the functional equivalent of the notion of property belonging to everyone; they both serve as the terms of a conversion process that results in the privatization of property" (Whitt 1995a: 9). The possibility of communal property rights is thus obviated by the assertion of individualized privilege.

To be sure, most Euro-American devotees of Indigenous spirituality would be appalled at the idea that their conduct is oppressive of the very people(s) they so admire. But, as we have seen, instrumental design is not always requisite to advanced colonial practices. Here again, it is the very fact of the appropriation, inseparable as it is from the obtrusion of property rights and the tacit subject-object dynamics of which it is simultaneously born and constitutive, which subverts the best intentions of the hobbyists and the spiritual sojourners alike. In much the same spirit as the early European settlers, they arrive uninvited to homestead on the cultural terrain of Indigenous North America. And, as it was in the case of so many missionaries and others who answered the consecrating call of the "White Man's Burden," a genuine—if misguided—altruistic intent is rendered inconsequential by the deleterious effects of their colonization.

But even if most of those who are attracted to Indigenous spiritual beliefs and practices seek in them a means by which to fulfill their own spiritual needs, still others have parlayed this attraction into a lucrative business. Here we encounter persons whose conduct is far more usefully understood as expropriation than as mere appropriation inasmuch as it relies on much more direct conjurings of property rights than the more subtle articulations implicit in the consumptive behavior of their followers.[23] And they have not as readily been afforded the presumption of sincerity sometimes granted their solicitous audiences. Lynn V. Andrews, for example, is the author of several commercially successful books and has been very much in demand on the New Age and pop-culture feminist lecture circuits. Though she is not an Indigenous person, Andrews is accepted by many as a legitimate and informed repository of teachings on traditional Indigenous knowledges and lifeways.[24] She is also frequently cited as a paragon of the commercial exploitation and distortion of Indigenous cultures and spirituality.[25] Andrews' "teacher" in her books, Agnes Whistling Elk, seems to the critics to be something of a collage: identified by Andrews as a Manitoban Cree, she has a Siouan name and draws on both Hopi and Lakota cultures in her teachings. To note this is not, of course, to denigrate or deny hybridity; it is not to insist upon the "pure," "pristine," or "authentic." But neither does the commitment to foregrounding liminality mean that we ought simply to take hybrids as found, treating them all as ontologically equivalent. Again, because hybridity is not unaffected by unequal power relations, we must take care not to lose sight of specificity in deference to it.

Returning to the particular, Jon Magnuson notes that, "two of the exotic ceremonies performed by Crees in [Andrews'] *Medicine Woman* are unknown among the Cree people of Manitoba, according to Flora Zaharia, former director of the Native Education Branch of the Manitoba Department of Education" (Magnuson 1989: 1086). Still, as Diane Bell points out, many of Andrews' readers are undaunted when confronted with inconsistencies in her accounts, insisting that, "true" or not, her musings are meaningful to them (Bell 1997: 52). But Indigenous critics of Andrews, Carlos Castaneda, and others, are rather less sanguine about the distorted accounts of Indigenous cultures and spirituality they accuse them of producing. Rose cautions that these accounts proliferate to the exclusion of Indigenous people's own accounts of themselves and their traditions (Rose 1992: 404). Emphatically underscoring this point, Whitt notes that "a leading figure of the New Age recently announced he intended to patent the sweat lodge ceremony since native people were no longer performing it correctly" (Whitt 1995a: 2). Similarly, Deloria expresses concern about the enduring negative effects of popular literature dealing with Indigenous people(s) more generally:

> . . . the book remains in the library where naive and uninformed people will read it for a decade to come. These people know only that the book

was once popular and assume that it has withstood all the challenges to its veracity that can be made. So they take the content of the book as proven and derive their knowledge of Indians from it. Over the long term, therefore, these books are extremely harmful to Indians because they perpetuate a mass of misinformation and improper interpretations for another generation of readers. Additionally, popular writers rarely return anything to the Indian community. (Deloria 1991: 459)

Confirming Deloria's fears, a quick check of the library catalogues of almost any major North American university is likely to find Andrews's and Castaneda's books shelved amidst more scholarly and autoethnographical treatments of Indigenous North American peoples.

As we have seen, Indigenous traditions lack an evangelical impulse to convert others—indeed, this would be anathema to many of them. With particular reference to literary appropriations of Indigenous voices, Silko draws our attention to a central and unresolvable contradiction inherent in whiteshamanism: "Ironically, as white poets attempt to cast-off their Anglo-American values, their Anglo-American origins, they violate a fundamental belief held by the tribal people they desire to emulate: they deny the truth; they deny their history, their very origins" (Silko 1979: 213). Thus, even the most prefatory gesture in the direction of culturo-spiritual appropriation/ expropriation is immediately and unavoidably untenable inasmuch as it violates values that are inseparable from the very traditions that the culturo-spiritual homesteaders purport to embrace. Again, an avowed right to choose for oneself and not to be bound by the objections of others, adjoined with an at least implicit rejection of the validity of communal rights, is central to the justificatory rhetoric of whiteshamanism. The resulting expositions of ostensibly Indigenous spiritual commitments, then, are necessarily disembodied and speak more as expressions of the persistence of the values and central tenets of Western cultural traditions.

A youth program offered by a New Hampshire polytechnic school usefully illustrates this tendency. Combining Lakota spiritual practices with Jungian psychology, the program engages middle-aged Euro-American men—calling themselves "elders"—to guide small groups of male students on weekend journeys of self-discovery.[26] The sweat lodge ritual, severed from its social and cultural contexts, is reduced to a means of personal fulfillment on these outings: part of an "initiation" into adulthood. In this respect, the program is not unlike many other New Age, Men's Movement, and pop-culture feminist performances of disembodied aspects of Indigenous spirituality. Cynthia Kasee describes such practices as part of a broader process of religious imperialism depriving Indigenous people and their communities of invaluable survival tools. As she puts it, "[t]his is *not* a victimless crime. . . . If Indian religions can be bought by any dilettante with a credit card, they lose their ability to require commitment, reform, and diminution of ego" (Kasee 1995: 83, 86; emphasis in original). In a similar vein, Churchill cautions that "to play at ritual potluck is to debase all spiritual traditions, voiding their internal coherence and leaving nothing usably sacrosanct as a cultural anchor for the peoples who conceived and developed them, and who have consequently organized their societies around them" (Churchill 1994: 213).

But might there be good reason to resist the invalidation of hobbyism and spiritual appropriation/expropriation to the extent that these phenomena and their attendant practices might be constitutive of a radical discourse of dissent in the dominating society? Certainly, much of whiteshamanism is premised on a rejection of at least some aspects of life in modern Western societies. However, in light of its distortions and their dire implications, the rendering of whiteshamanism as oppositional voice is not sustainable. The commodification so typical of the New Age, Men's, and pop-culture feminist movements, for example, being bound up with more pervasive processes of advanced colonialism, leaves little room for any credible claim to the contrary. As Whitt argues:

> When the spiritual knowledge, rituals, and objects of historically subordinated cultures are transformed into commodities, economic and political power merge to produce cultural imperialism. A form of oppression exerted by a dominant society upon other cultures, and typically a source of economic profit, cultural imperialism secures and deepens the subordinated status of those cultures. In the case of indigenous cultures, it undermines their integrity and distinctiveness, assimilating them to the dominant culture by seizing and processing vital cultural resources, then remaking them in the image and marketplaces of the dominant culture. (Whitt 1995a: 3)

The appropriation/expropriation of Indigenous culture and spirituality thus threatens to subvert the resistances already embodied in those traditions. In so doing, it buttresses enduring structures of advanced colonial domination. Far from constituting a radical discourse, then, whiteshamanism manifests as a potent—if not necessarily instrumentally conceived—practice in support of advanced colonialism.

Even the otherwise emancipatory discourses of pop-culture feminism have this effect. To be sure, discursive and semiotic appropriations from hegemonic groups by marginalized ones may have subversive, even transformative potential—as in what Bhabha calls mimicry, for example. But appropriations from more marginalized groups by the comparatively privileged, even if they may appear on the face to be subversive in relation to still larger hegemonies, are fated to further entrench the subordination and voicelessness of the people being imitated. Of course, we may well imagine real possibilities for the building of alliances between feminists and Indigenous people(s), but these will have to be based on efforts to create a space in which marginalized voices can be heard rather than on the presumption that those voices can be raised through self-appointed surrogates.[27] Pop-culture feminist appropriations, far from redressing the voicelessness of the peoples upon whose traditions they draw, perfect it to an extraordinary degree by usurping even the authenticity of Indigenous voices. Owing to their relative privilege, reflected most conspicuously in the ability to publish, Andrews and other pop-culture feminist personalities enjoy much greater discursive reach in the dominating society than do most members of Indigenous communities. Moreover, as Laura Donaldson points out, the individualized, self-help orientation of Andrews's brand of "feminism" runs contrary to the ethical context of the culturo-spiritual traditions upon which she draws (Donaldson 1999: 692–93). Here again, a privileging of individual rights over collective responsibility is betrayed—in this case, as Donaldson argues, the promise of emancipation for Euro-American women is made at the expense of whole Indigenous communities. Thus, while pop-culture feminist appropriations may be imbued with emancipatory potential, it is a circumscribed potential that attends to the well-being of Euro-American women at the expense of more marginalized people(s).

The conduct of the hobbyists also does not escape implication in this. Churchill describes repeated encounters with hobbyist groups while on a speaking tour in Germany, and notes with some sense of irony how their members readily expressed their "collective revulsion to the European heritage of colonization and genocide" (Churchill 1994: 224). By itself, this is certainly a praiseworthy sentiment. But when explicitly connected to the recreational assumption of a counterfeit Indigenous identity, it becomes rather more problematic and infinitely less compelling. Whatever latent counter-hegemonic potential may reside in the disembodied symbols and rituals adopted by the hobbyists is squandered when they are grafted onto the hegemonic structures of the dominating society. In Churchill's view:

> The upshot of German hobbyism . . . is that, far from constituting the sort of radical divorce from Germanic context its adherents assert, part-time impersonation of American Indians represents a means through

which they can psychologically reconcile themselves to it. By pretending to be what they are not—and in fact can never be, because the objects of their fantasies have never existed in real life—the hobbyists are freed to be what they are (but deny), and to "feel good about themselves" in the process. And, since this sophistry allows them to contend in all apparent seriousness that they are somehow entirely separate from the oppressive status quo upon which they depend, and which their "real world" occupations do so much to make possible, they thereby absolve themselves of any obligation whatsoever to materially confront it (and thence themselves). (Churchill 1994: 225)

In this sense, then, the expropriation of Indigenous culture and spirituality furnishes something of a salve for open(ing) sores on the dominating society and is thereby converted into a conservative force for system maintenance to the extent that it rechannels discontent that might otherwise be harnessed as dissent in furtherance of change. Moreover, since the status quo of power relations ensconces advanced colonialism, Indigenous peoples' own cultural and spiritual traditions have thus arguably been turned against them.

An apt metaphor for the culturo-spiritual expropriations of whiteshamanism is the particular form of alchemy developed and practiced in Europe during the Middle Ages. Simultaneously a philosophical movement and a practical endeavor, alchemy held out the dual promise of enlightenment and enrichment. Believing nature to be unequal to the task of perfecting itself, the alchemists sought to reveal the hidden patterns in the natural world and, by means of manipulation, to urge them out into the open in the hope that they could be harnessed and turned to more utilitarian purposes. In a similar spirit, the would-be recipients and beneficiaries of traditional Indigenous knowledges seek to uncover new ways of knowing and being in the world, and presume by means of manipulation—resulting, as we have seen, in distortions—to make them relevant to their own lives. And just as some alchemists pursued the fabled philosopher's stone solely for the pure truth and enlightenment that it was believed to embody while others coveted it only as a means to turn base metals into gold, so too the ranks of whiteshamanism include both genuine seekers of spiritual fulfillment and those who see Indigenous cultures and spirituality as vehicles for the promotion of their own material indulgences.

But the new culturo-spiritual alchemists are engaged in far more destructive and, often, self-deluding practices. Unlike their metaphorical antecedents, they actually threaten by their ill-conceived antics to destroy precisely that which they seek to exploit. The often absurd compounds of the conjuring alchemists of yore never imperiled either the existence or the value of real gold. Whiteshamanism, however, bears a range of pernicious effects that both devalue and debase Indigenous cultural and spiritual traditions while, in significant ways, jeopardizing the survival conditions for real Indigenous people. Among these are the powerful and sustained contribution they make to the objectification of Indigenous people(s), the direct and indirect support they give to the structures of advanced colonialism, and, perhaps most damaging, the culture change that accompanies their endeavors. This is not to say, of course, that culture change would not—or even should not—occur in the absence of whiteshamanism. Black Elk's catechism, for example, was a source of some degree of such change, even if we accept Rice's interpretation of it. But clearly there is a significant difference between culture change produced from within by Black Elk behaving like a buffalo and that which, sustained by unequal power relations, is induced from without by the unscrupulous and the naïve in sites of comparative privilege.

The varied manifestations of whiteshamanism cannot be examined in isolation from the more direct forms of ethnocide and genocide perpetrated against Indigenous peoples over the last several centuries. Nor can they be marked out as discrete and separate from the ongoing structures and processes of advanced colonialism. This stands as a powerful caution against treating hybridity such that it might be seen to naturalize the enduring distortions of whiteshamanism. Apart from treading dangerously close to evolutionism or a decidedly Western notion of "progress" this leaves precious little basis for an understanding of the often strong reactions of Indigenous people(s) against what they regard as the misuse and abuse of their traditions. If the distortive effects of whiteshamanism are entirely normal, natural, and inevitable, then the vehement protestations against them can hardly be cast as other than bewildering and disproportionate, even irrational—a perspective that is immediately reminiscent of the politically disenabling rendering of the Grass Roots *Oyate* as self-deluding. And it is precisely this unsatisfactory implication that signals the need to explore the broader contexts of unequal power relations in which the issue of whiteshamanism is nested.

What this suggests is that it is not enough simply to condemn the conduct of hobbyists or New Agers as problematic for its distortive effects. As suggested by Judith Butler's work on hate speech,[28] this runs the risk of reducing the problem to the agency of the subjects whose (mis)appropriations sustain these violences, thereby missing the broader processes of advanced colonial *consumptive domination* that make their actions meaningful. Put another way, it is the terms upon which such hybridizations are founded and the conditions that make them possible—in contrast to other imaginable appropriations that might, as Hernández-Ávila suggests, be less tenable—that are most deserving of our attention. Although my description of whiteshamanism has required an elaboration of details in ways that seem to foreground its agents, it is their mutual implication in/as practices embedded in the broader processes of advanced colonialism that are important. What this should signal is the imperative that we always make the interrogation of hybridity central to our analyses; that is, we should take care to acknowledge that hybridity is inherently the product of negotiation between heretofore discrete ways of being and that, when carried out between sites of relative margin and privilege, negotiation not infrequently produces winners and losers.

For Bhabha, as we have seen, hybridity is an essential consequence of trying to replicate texts in different contexts—something is always changed by the context in which a text is being reproduced. But if this is the case, it must work both ways. This means that if hobbyists, New Agers, or some feminists try to reproduce Indigenous texts in the context of the dominating society, hybridity must result here too. The problem is that the hybrids made in the dominating society have the material-discursive backings of its technologies, its hegemonies. And this threatens to extinguish the original texts—something that Indian hybrids of English texts under the Raj, for example, never threatened to do. That is, unlike those of colonial India, the hybrids here begin to aver being "the real thing" and, what is more, their site of hegemonic privilege confers the power to sustain that pretense. Extending the alchemy metaphor, they advance a claim to be gold and validate that claim, in part, by substituting themselves for actual gold—for all intents and purposes, then, they become gold. Appropriated spirituality thus becomes that which it seeks to emulate not by replicating it, but by displacing it and usurping the spaces it once occupied: it is thus that some of Andrews' followers might miss the absurdity of their chastising an Indigenous woman for not knowing herself.

As we have seen, Bhabha treats mimicry as a resistance or survival tactic of the colonized that unsettles the stable categories of oppositional identity by forcing the colonizer to recognize elements of the Self in the colonized Other. Resistance is located in the resulting subversion of hegemonic narratives and ideas. Similarly, Appadurai describes how elements of American culture are "indigenized" as and through resistance to globalization (Appadurai 1996: 29–33). The importance of this is that, like Bhabha's mimicry, it places resistance alongside domination/dissemination, reminding us that the rest of the world is not reducible to passive skin awaiting inscription by (in this case) globalizing American culture flows. This, in turn, lays bare the site of a political praxis and, with it, the agency of the Other.

But as the phenomenon of whiteshamanism suggests, mimicry (like hybridity more generally) is not read only from experiences of the colonized.

On the contrary, the colonizers too are sometimes found to be engaged in mimicry of their own, affecting simulations derived from their Others. As we have seen, however, these appropriations from marginality can also be much more than mere simulation. Their performance being wholly inseparable from the unequal material and power relations of advanced colonialism, they become more in the way of simulacra, though this still does not quite capture the essence of the process. As in Jean Baudrillard's treatment of media coverage of the 1991 Gulf War, the simulacrum is a copy that exists without the original, occluding the nonexistence of the truth it ostensibly represents so that, in effect, it is that truth.[29] But something qualitatively different is discernible in mimicry of the colonized by the colonizer. Here the occlusive properties of the simulacrum merge with the (re)inscriptive capacity of the *coup de force* in a phenomenon that is neither: the whiteshamans *become* the object of their own solicitous desire in the instant that the violences of their appropriations banish it. This is *alchemy*—though perhaps emancipatory in conception, it is the ultimately conservative "opposite number" to the potentially emancipatory mimicry described by Bhabha.

It is the combination of appropriation and the possibility of a founding in emancipatory hopes that makes alchemy a unique phenomenon. Generative of hybrid identities/sites in much the same way that mimicry is, it connotes something more than mere (re)presentation. In particular, it is distinct from those discursive representations, discussed in earlier chapters, that enable the self-knowledges of the dominating society through *différance*. Indeed, its conjurings tend, at least implicitly, to be set in radical opposition to hegemonic narratives and ideas. Alchemy, then, unlike some of the more subtle workings of *différance*, proceeds from a necessary instrumentality that is explicitly programmatic, involving practices of appropriation that are consciously conceived. Even where the motives of the alchemists might seem cynically self-indulgent, however, this instrumentality cannot be read as being consciously in furtherance of the violences of advanced colonialism. But the hegemonologue works through alchemy nonetheless in ways that are no less implicated in advanced colonialism for being unintended (or even contrary to what might have been intended): in the liberal-individualist underpinnings of appropriation, for example.

None of this is to suggest that a retreat from notions of hybridity and mimicry—and all that they enable—is at all warranted, much less desirable. On the contrary, these important contributions from recent postcolonial theory are invaluable for having highlighted the complex negotiation of subjectivities in the contact zone and for recovering agency and resistance from inscription and desire. But, as suggested by what Hale tells us about hegemonic appropriations of the idea of hybridity, we pay too high a price for all of this if we do not at the same time remain sensitive to specificity. Similarly, a nuanced account of hybridity itself must take account of how sites of privilege and margin effect the negotiation and construction of identity. Bhabha's insights about mimicry might thus be complemented by a fuller understanding of what I am calling alchemy: if the former reflects what Appadurai calls the "indigenizing" of the hegemonic, then we might say that the latter indicates a "hegemonizing" of the indigenous/marginal (undercutting resistances in the process). Crucially, this also foregrounds the relativity of margin and privilege, thereby demystifying the violences arising from practices—whiteshamanism, for example—that might otherwise seem imbued with emancipatory potential.

Toward Counter-Hegemonic Critique

Universalized concepts like individualism, secularism, class, and gender have given rise to emancipatory projects that, in their essence, dictate universalized terms of emancipation, imposing them tyrannically without sufficient regard to context and in defiance of alterity. When emancipatory theories seize on ideas such as these, insisting upon them in ways that enable the articulation of some particular vision of "the good life," they repeat the ethnographer's mistake of presuming to speak on behalf (and in place) of the Other. Likewise, the more general inability to step outside of the cosmologically defined contexts of the dominating society—whether in the productive sense of proposing the terms of emancipation or in the more deconstructive sense of deep critique—continues to privilege the hegemonologue. It is thus that emancipatory theories share in the orthodoxy's complicity in the ideational dimension of ongoing advanced colonialism.

It is also important to note that emancipatory theory in general remains mired in the disenabling conventions by which we authorize knowledge in the dominating society. Developed and articulated through the academy and the structures of knowledge production authorized and sustained by the dominating knowledge system, emancipatory theories are limited (and limiting) in that they can achieve no more than a contra-hegemonic stance. The problem as I have described it is that they participate in that game of hegemony's making: contending against one another in a way that reproduces the logics of binary opposition. That is, the articulation of different approaches takes place on the oppositional terrain of the academy such that they ultimately contest for hegemonic position, rather than against it. Extant emancipatory international theory is thus "sympathetic" in its orientation to cosmologically distinct

Others, but falls well short of the empathetic ideal. As I have argued, *différance* and the prismic distortions of intersubjectivity ensure that true empathy—in the sense of seeing through the eyes of the Other—can never be realized. As an ideal, however, this very impossibility enjoins us to affirm a commitment to conversation and thus to admit of cosmological possibilities that are radically different from our own—an approach that goes beyond the mere acknowledgment of the authority of marginalized voices, necessitating that we hear them directly instead of through surrogates. For want of conversation and deference to the empathetic ideal the matrices of the intertext cannot be traced, leaving us ill equipped to confront our complicities with the hegemonologue and, through it, with advanced colonialism. Inasmuch as the various emancipatory approaches to theorizing the international have failed in this regard, International Relations has yet to see truly *counter*-hegemonic theory.

All of this is also quite revealing of the considerable extent to which International Relations has itself suffered violences traceable to the colonial experience. As noted in earlier chapters, the postcolonial subject of interest here is not only the Indigenous peoples marginalized by the corpus of international theory, but international theory itself which has been both shaped and constrained—in short, subjugated—by colonialism. Before we can entertain hopes of developing broadly emancipatory international theory, then, international theory itself must be emancipated. And this means, above all, addressing the workings of a larger social imaginary that has hitherto left us content to exclude Indigenous peoples from international theory or, at best, to presume the sufficiency of our own voices in their stead. This is a vital first step toward the imperative that the hermeneutic circle be rehabilitated as an unabbreviated polyphonic site of sustained conversation. In the meantime, the incompleteness of the intertext demands that we disabuse ourselves of any pretension to having mounted a broadly counter-hegemonic project in disciplinary International Relations. To pretend otherwise is to participate all the more fully in the hegemonologue, denying its violent erasures and, with them, our own. And as my own call for conversation still issues from a preconversational moment, the best that can be claimed is that I hope to be working *toward* a counter-hegemonic critique at which we have yet to arrive.

Notes

1. Poststructuralists, for example, do not in every instance venture from deep critique to articulate their own accounts of "the good life." Indeed, this is not uncommon where "critical" theories are concerned inasmuch as they must often work first to

open "spaces" in which to consider new possibilities through critique of dominant perspectives. Still, the very founding of social critique, unless it is wholly gratuitous, necessarily proceeds from some vision of a more desirable order. And as I discussed at length in chapter 1, the intense solipsism sometimes attributed to poststructuralism—and claimed as evidence that it defies a politics—betrays either a misuse of a Foucault, Derrida, or Irigaray or a failure to read through their works.

2. Though Wight spoke as to the whole of the discipline, he did so from within the discursive world of Realism, taking its assumptions as a true reflection both of the international and of International Relations itself.

3. According to Anne Marie Goetz, feminist political involvement in issues associated with women and development "vividly illustrated contradictions between a feminist theory engaged in the dismantling of oppressive categories of social organisation and a politics which is predicated on the unity of the (universalising) category of 'women' " Goetz (1988: 477).

4. See, e.g., Lugones and Spelman (1983).

5. It is perhaps thus that, as discussed in chapter 1, some Indigenous critics of "postcolonialism" have raised the objection that direct colonialism endures to this day. Even in literature, the "postcolonial condition" cannot be read as post-colonial. As Lee Maracle puts it: "Unless I was sleeping during the revolution, we have not had a change in our condition, at least not the Indigenous people of this land. Post-colonialism presumes we have resolved the colonial condition, at least in the field of literature. Even here we are still a classic colony. Our words, our sense and use of language are not judged by the standards set by the poetry and stories we create. They are judged by the standards set by others" (Maracle 1992: 13).

6. As Locke put it, "An acre of land that bears here twenty bushels of wheat, and another in America, which, with the same husbandry, would do the like, are, without doubt, of the same natural, intrinsic value. But yet the benefit mankind receives from one in a year is worth five pounds, and the other possibly not worth a penny; if all the profits an Indian received from it were to be valued and sold here, at least I may truly say, not one thousandth" (Locke 1924: 137).

7. See, e.g., Oneal and Russett (1999).

8. See, e.g., Fukuyama (1992). For a formulation that, among other things, equates the global proliferation of McDonald's restaurants with the spread of liberal values requisite to sustain peace, see Friedman (1999).

9. Particularly noteworthy in this regard is Lenin's account of imperialism wherein he holds that "the colonial policy of the capitalist countries has *completed* the seizure of the unoccupied territories on our planet" Lenin (1939: 76; emphasis in original). The idea that the territories seized by the colonial powers had been "unoccupied" is, of course, the functional equivalent of Locke's pronouncement that "in the beginning, all the world was America."

10. See Engels (1972).

11. The foundation of Engels's "primitive communism" was, through Marx, derived from propositions by Lewis Henry Morgan generalized from his studies of the

Iroquois. See Morgan (1877). Morgan's account of progressive development from "savagery" to "civilization" has been roundly criticized for its social evolutionism—an orientation that, as we have seen, has been deeply implicated in the violences of colonialism/advanced colonialism. Nevertheless, its echoes can be heard even in more contemporary international theory. See, e.g., Cox (1987: 36–38).

12. See Black Elk (1983: 148–50).

13. The apparent inability of Western scholars to come to terms with this idea is perhaps best reflected in their problematic efforts at taxonomy. The most common appellation, "berdache," is both misleading and culturally inappropriate; the designation "man-woman" collapses the independent ontological significance of liminality into the poles of a rigidly dichotomous rendering of gender that remains very much intact.

14. Although instances of social ridicule have been cited by some, these have turned out to have had sources other than the transgression of gender norms. See Greenberg (1985). The disciplining practices of the dominating society have, since the colonial encounter, worked to quite great effect to force *winkte* into concealment. See Callender and Kochems (1983); Williams (1985).

15. See Schnarch (1992); Thayer (1980).

16. See Callender and Kochems (1983).

17. On the implantation of Western gender roles, see Medicine (1981); in a similar vein, see Notarianni (1996).

18. Similar failings underwrite a wide range of misinterpretations of Indigenous social practices. As Marla N. Powers (1980) points out, for example, women's seclusion during menstruation reflects reverence for sacred powers of reproduction believed to inhere in women, not a form of banishment.

19. See Tohe (2000).

20. These are "de facto" boundaries in the sense that they are not legislated in place by the terms of the theory itself. Rather, they arise in consequence of the failure to establish a fuller intertextuality through a commitment to conversation.

21. As argued in chapter 3, the violences of misinterpretation or decontextualization result ultimately in erasure as the voices of ethnographic subjects-cum-objects are displaced by those of their ethnographers—a process by which Indigenous knowledges are inevitably made to accord with the predetermined concepts and categories of the dominating society, serving thereby to confirm all that was already "known." Nothing could be more instructive in this regard than an idea like "primitive communism" or the (mis)readings of Lakota gender roles.

22. Elizabeth Neuffer, "Germans Make a Hobby Out of Cowboys and Indians," *Boston Globe* (August 6, 1996), p.E1. Although hobbyism exists in North America as well, it has enjoyed its greatest popularity in Europe. And European hobbyism is particularly instructive for us here inasmuch as it underscores that the violences of advanced colonialism need not be performed locally in order to have local effects through the hegemonologue.

23. In a related context, Churchill also makes a distinction between appropriation and expropriation, linking the latter to self-congratulatory claims emanating from within

the Men's Movement to the effect that its adherents have displaced Indigenous people as the authentic practitioners of their traditions (Churchill 1994: 217–18).

24. Recall that it was Andrews's followers who, when she was challenged by a Cree woman disputing her accounts, rudely insisted that their mentor knew better the "inner meaning" of Indigenous culture. Andrews is perhaps best known for her *Medicine Woman* (1981).

25. See, e.g., Castile (1996); Jocks (1996); Kehoe (1990); Rose (1992); Whitt (1995b).

26. See Gose (1996).

27. This is not to conflate the advanced colonial complicities of pop-culture feminisms and feminist scholarship. The latter are treated earlier; the former are invoked here only as one example in support of the caveat I wish to raise on the question of hybridity.

28. Butler's concern is that prosecutorial-judicial solutions to the matter of hate crimes, becoming engaged only on a case-by-case basis, in effect conflate the problem with its symptoms. Thus, Butler argues, individual instances of proscribed behaviors eclipse the broader workings of racism that enable them (Butler 1997).

29. See Baudrillard (1995).

PART 3

Reflection

CHAPTER 8

Conclusion: Recovering International Relations from Colonial Practice

In the preceding pages, I have argued that international theory, indeed social theory more generally, is cosmologically inflected in ways that (re)produce the restrictive hegemonic concepts, categories, and commitments of the dominating society. I have pointed out that there are important senses in which it is monological and spoken from positions of conspicuous privilege. I have gone so far as to claim that all theory is violent. And yet I do not propose that we either can or should want to dispense with theorizing, for to do so would be to accept a definition of theory as something it need not be. Bhabha draws our attention to what he describes as the "damaging and self-defeating assumption that theory is necessarily the elite language of the socially and culturally privileged" (Bhabha 1994: 19). It is nothing of the sort. It might, however, be fairly claimed that it has been appropriated by the socially and culturally privileged to the extent that it is they who have defined what counts as theory and what does not. It is thus that theory has mapped comfortably with the voice of the hegemonologue—a voice that, through the authority of its scholarly raconteurs, has had the audacity to hold that it, and it alone, speaks bona fide theory. Impoverishing International Relations, this pretension has underwritten the confinement of Indigenous people(s) to other disciplinary contexts as well as their treatment as objects of study rather than as credentialed speaking subjects in their own right.

This is at the root of our failure, thus far, to develop truly *counter*-hegemonic theory in International Relations. Though there is, to be sure, a wide range of valuable critical approaches in our existing theoretical repertoire, these turn out to be sympathetic—if not in conception, then in application[1]—falling short of the empathetic ideal as I have described it. What this suggests in the first instance is that we need to move beyond the usual predisposition of emancipatory discourses to dictate the terms of emancipation since, as we have seen, this runs the considerable risk of substituting one form of oppression for another. It therefore calls upon us to rethink the very fundamentals of the emancipatory project, dispensing with the paternalistic predisposition to define and legislate the terms of emancipation. It is tempting, then, to propose that empowerment ought to be our true objective. But this resets the trap, for who will decide what constitutes power and its sources and whose sense of the desired ends will prevail? No, emancipation can be born only of enabling moves that come from an opening up of possibilities—not by pursuing a particular programmatic agenda, but by working to unsettle the hegemonic structures, narratives, and ideas that prevent a fuller range of potentialities from being enacted or even imagined in the first place.

All of this will come to naught, however, unless advanced in tandem with a sincere commitment to conversation. Having laid bare the monological core of disciplinary International Relations, the task ahead is to work for polyphony. But this does not mean that we confer upon ourselves the responsibility to bring Other voices into the discipline. Lacking the competencies to speak for Others, this would be to miss the point entirely. Moreover, these are not competencies that we should hope to acquire for ourselves in the first place since, as I have argued, the oft-prescribed means by which to do so, participant-observation, is the antithesis of conversation—in raising a surrogate voice to speak in place of the Other, the Other's perpetual outsider status is reconfirmed. In any event, such knowledges as we might (imperfectly) apprehend would be doubly suspect for having been disembodied from their broader sociopolitical and cultural contexts. One need only recall the fate of cross-disciplinary appropriations of knowledge—ideas frozen into stasis in defiance of new insights and developments whence they came—to imagine the pathological effects of this. All of this serves to remind us that empathy can, for the theorist, never be more than an ideal. However, we can still hope to construct theory that is empathetically inclined, let us say, to the extent that it readily admits of cosmological possibilities that are radically different from our own—this would be theory that not only admits of marginalized voices but that also necessitates that they be engaged in conversation. Conversation, it should be emphasized, is an empathetic commitment; it is also the first indispensable counter-hegemonic move.

The importance of this is perhaps best illustrated by an irony I observed at the outset of this book: that I would construct a linear argument, in seeming defiance of some of the very points I wanted to make. And so, progressing in linear fashion, I have traced a connecting line, moving from one to the next, through core conceptual commitments, disciplinarity, ethnography, popular culture, orthodox social theory, and emancipatory theories. But I would propose now that, even if this has, for the sake of expediency, been the way I have proceeded, we have arrived at a vantage point from which we can see that our travels have taken us less along a continuum than through a complex. Saying this is not to homogenize Western voices and experiences. There is, however, a discernible recurrence in the themes taken up in each of the preceding chapters, which is itself a demonstration of what I have been arguing. Take, for example, the apparent inability of some critics of poststructuralism to recognize that the grounds upon which they issue their indictments are among the very things called into question by the theorists of whom they disapprove. This failure to read through other theories on their own terms is illustrative of the broader problem of egocentrism in translation and the violences of inscription that so often follow from it; a cautionary tale about the consequences of unreflexively reading through hegemonic referents. In this sense, the discussion of the Sokal hoax in chapter 1 is proffered as more than just a preemptive response to those who might charge ineluctable relativism in reply to some of my theoretical commitments—it is simultaneously a prefatory instance of a problematic found to reside in a range of other contexts explored in subsequent chapters.

A salient aspect of the misrenderings of poststructuralism, however, is that they have not gone unanswered. Even if those to whom the replies have been addressed might not always have heard them, others certainly have. Indeed, there is a substantial literature that, like Ashley and Walker's "dissident" thought in International Relations, has been quite influential both in itself and in terms of opening a space for new voices of dissent. And though engagements with these voices by scholars with more "mainstream" inclinations are still too infrequent, the very fact of Sokal's hoax and the extent of controversy it stirred makes clear that it is increasingly difficult to simply ignore them. But the same cannot be said of those issues and ideas that, though they may be relevant to work being done by students and scholars of International Relations, have effectively been concealed from them by disciplinary practices. Nor is the inaudibility of voices spoken elsewhere in the academy meaningfully disturbed by interdisciplinary forays that, as we have seen, typically amount to little more than moments of appropriation and seldom result in sustained conversation. One consequence of this is the prospect of an immediate (by way of misapprehension/misrendering) or longer-term (as ideas fall to challenges and are discredited in their original disciplinary contexts) disconnect between the state of knowledge in the discipline of origin and how that knowledge is represented and engaged by those with other disciplinary affiliations. The problem here is analogous to the misrenderings of poststructuralism, but is worsened by its potential to go completely unanswered.

More troubling are circumstances wherein the inaudibility of those who would answer our (re)presentations owes to unequal power relations. It is lamentable that some scholarly voices sometimes have greater discursive reach than others, particularly than those at the margins. And yet, the "margins" of the academy are still sites of conspicuous privilege populated by voices that carry a good measure of authority for having been credentialed. Even across the reified boundaries of the rarified disciplines, the possibility at least exists for extradisciplinary voices to speak authoritatively in answer to the deployment of appropriated knowledges. But voices differently credentialed are another matter altogether. In the case of the people(s) "without history" voice itself is appropriated. Underwritten by the colonial complicities of the disciplinary division of knowledge, the possibility of an exogenously sited speaking position is obviated by the pretensions of participant-observation and ethnographic representation that, together, reduce Other subjectivities to contained objects. Exacerbated by profound inequalities of power and authority/reach of voice, a problem similar to that first highlighted in Sokal's treatment of poststructuralism is here identifiable with and as something more: the hegemonologue. Occluding possibilities of which its cosmological boundedness will not admit, it is spoken as clearly in well-intentioned pop-culture treatments of Indigenous peoples as in those that self-consciously allied themselves with the worst excesses of colonial violence. Likewise, it is telling that its erasures are found to be carried forward in the orthodox theoretical commitments of scholars who might be surprised to discover that their work has any connection to knowledges about Indigenous peoples and, to varying degrees, can inhere in explicitly emancipatory designs as well.

What this palpable recurrence signals is that the strategic linearity of this book has not traced merely a series of disparate points, but a single knowledge-complex defined by colonialism and delimited by cosmology. It is, in essence, a story made up of and spoken through a multiplicity of stories whose interconnectedness we have not well enough appreciated. International theory is woven through and fashioned out of the commitments that make up this knowledge-complex. Of course, these observations might prompt the objection that what these explorations have led me to advance is not science, but allegory.

But none of what has been said herein is unsettled by this charge unless it is imagined that there are non-allegorical tales to be told in and of International Relations. The close ontological ties that, to take one example, bind orthodox international theory to the theoretical orthodoxy of Anthropology belie the former's pretense to science. Far from representing a triumph of science over allegory, these orthodoxies have effected a "sciencing-over" of allegory, cloaking unsubstantiated ideas like Hobbesian-inspired accounts of human nature in the garb of epistemological rigor and certain knowledge. Thus, while I freely concede that what I have advanced herein is allegorical, so too are the dominant theoretical discourses of International Relations.

In answer to these dominant discourses it is tempting, if only by way of conclusion, to suggest something of how the key concepts and categories of International Relations might be differently imagined from a traditional Lakota perspective. In our as yet pre-conversational moment, however, venturing to do so would inevitably work new violences without the possibility of their being answered. The problem we face is that International Relations must also be recovered from colonial practice. Importantly, then, a point I would hope to advance most emphatically is that my own admittedly allegorical narrative does not speak only to the advanced colonial violences visited upon Indigenous people(s). That is, conversation is not advocated in the interest of the Other alone. The academic pursuit of knowledge stands to reap tangible rewards of its own if recovered from colonial practice. Not only does the possibility of broadly counter-hegemonic theory and practice depend on it, but as postcolonial subjects ourselves we also stand to learn something of, as Geertz puts it, those minds we find ourselves to have.

As suggested in chapter 7, the creation of a space in which marginalized voices can be heard and the recognition that those voices deserve an answer opens up the possibility that we might forge alliances. Here, then, is the basis for a truly emancipatory project. Illustrative of the promise in all of this is a cooperative effort undertaken by a small group made up of archaeologists and holders of Northern Cheyenne oral history, with the aim of reexamining the circumstances of a much-mythologized moment in the history of the American West: the breakout from custody at Fort Robinson, Nebraska by Dull Knife's Northern Cheyenne band on January 9, 1879.[2] Together, the members of this collaborative research team undertook to settle a major point of disagreement between dominant Euro-American historiographies of the event, on the one hand, and much lesser-known accounts from Cheyenne oral tradition on the other. The dispute was of more than passing importance to the Northern Cheyenne members of the group for two main reasons: besides the fact that the reliability of oral history was clearly at stake, the Euro-American telling of the breakout story cast Dull Knife and his band in a rather poor light, holding that they had attempted to make their escape from the pursuing soldiers along a route that would have been nothing short of foolhardy. The researchers, however, were able to uncover clear evidence in support of the Northern Cheyenne account—corroboration that was forthcoming only because the advices of oral history led the archaeologists to find things they might never have found in places they would not otherwise have looked. This, of course, underscores that the outcomes of scholarly knowledge production are dependent upon the questions asked and, even more fundamentally, on the starting assumptions that made it possible to develop those questions in the first place. "History and its auxiliary science of archaeology," the investigators later observed, "become part of the architecture by which people are dominated" (McDonald et al. 1991: 75). As their own experience demonstrates, however, this does not preclude deploying History and Archaeology—or International Relations for that matter—in furtherance of resistance, provided they are first disabused of their monological pretensions.[3]

But despite the success of such laudable collaboration and what it tells us about the value of oral literatures, it is still not enough just to listen to other voices because there is no way of divorcing that exercise from the power relations of the colonial encounter and ongoing advanced colonialism. Moreover, to the extent that empathy can be no more than an ideal, we should not delude ourselves into thinking that simply by acknowledging the hegemonologue and the unequal relations it works to sustain we might also elude complicity in advanced colonial violences. I do not conclude the present project, then, with the line of critique that has been developed through the preceding chapters. Nor, as I have said, will I issue a call to bring Indigenous peoples' issues into International Relations. Indeed, it turns out that Indigenous peoples have never really been absent from international theory after all—essential knowledges about them have been taken up and (re)produced by the orthodoxy (with special emphasis on their presumed "nature") and critical approaches (primarily in the ascription of some necessary emancipatory agenda, but also by dint of cosmological myopia) alike, even if they have seldom been made explicit. Given their indeterminacy, these are certainly knowledges that cry out to be destabilized, but that does not complete the project.

Of course, though we might be able to say that knowledges *about* Indigenous peoples have always been part of international theory—most fundamentally through *différance*—the same clearly cannot be said of Indigenous peoples' *own* knowledges. And yet, I resist calling on scholars working in any of the existing traditions of international theory to inform their work with reference to Indigenous cosmologies. This would bespeak appropriation that, as we have seen, leads inevitably to new violences of ascription and/or erasure. The problem here is in the framing of the project—that is, in the hope that any existing approach from the corpus of international theory might unproblematically draw on Indigenous knowledges. The distinctive worldview and ways of knowing of Lakota traditionalism, for example, do not at all lend well to any of our existing discourses on notions like security or, more broadly, of the international. Two main lines of pathology are immediately discernible in any imaginable suggestion to the contrary: first, the operant cosmology of Western academic discourse works violences upon commitments and ideas "otherwise constituted" when they are forced into the confines of the possibilities it defines (and denies); and, the very pretension to appropriate knowledges and synthesize them into "theory" signals a prior assumption of a hierarchical ordering of authoritative voices, reconfirming that of the scholar while once again reducing Indigenous people themselves to ethnographic subjects-cum-objects. Even sincere emancipatory designs are subverted here by the mistake of misconceiving our responsibility as being *for* rather than *to* the Other.

Again, it is only in conversation that relative authority and authenticity of voice is confirmed. What does it mean, then, for us to take Indigenous voices seriously as conversational partners when we do social theory? Most fundamentally, it means recognizing that Indigenous people(s) and their knowledges should be of interest not because we might suppose that they can inform our theories, but because what they have to tell us *is* bona fide international theory in its own right. This is Bhabha's point in rejecting as self-defeating the assumption that theory belongs to the socially and culturally privileged. We are thus enjoined to engage in conversation in the fullest sense, acknowledging that those voices we would seek to make audible await more than the opportunity to be heard. They are also entitled to hear back from us, just as any other articulation of international theory is deserving of a considered response. This is, after all, no less than what those of us who work at the margins of International Relations have insisted upon where our own work is concerned. To converse, then, is first to presume a mutual authority to speak and obligation to respond. To do otherwise is to accept in some measure the theory/superstition binary and all the deferred meaning it entails. Here, then, is a prerequisite for any counter-hegemonic project.

Taking Lakota cosmology as constitutive of authentic international theory begs the question of what its distinctive contribution might be. Surely, the answer must lie in its ability to treat alterity without constructing hierarchy—it is a discourse that allows us to find hope for peace in difference rather than *in spite* of it. It does not automatically promise peace—conflict, as we have seen, was a part of the aboriginal condition—but it does naturalize it, taking conflict as the exceptional case. And it does this without working to transcend difference and, therefore, without conflating the violences of erasure with the ends of emancipation. But is Lakota cosmology as neatly parceled and focused as some might expect international theory to be? While it is a discourse of the international, it is much more than that too. Dissenters, especially those working in the orthodox theoretical tradition, might thus complain that, owing to its holism, it is a much more expansive and unwieldy thing than a theory of the international ought to be. But, as I have endeavored to show, there is much more than a theory of the international bound up in orthodox international theory too—voices from the travelogues, the idea of original sin, and much more can be found just beneath the scientific veneer of Realist-inspired theory, for example. This reveals that our own accounts/discourses of the international are infused by the whole of our social experience as well. Fundamentally, this is because International Relations has only been artificially marked apart from a much wider terrain populated by knowledges it has resisted calling its own. When we pull back the curtain of disciplinarity, this broader foundation, we find, is one of the things it conceals.

It is in this limited sense that we are able to talk about the intertextuality of international theory. But it remains an intertextual space with discernible boundaries. Western cosmology—spoken through the hegemonologue, itself fixing limits upon the popular imaginary of the dominating society—has dictated exclusive terms of admittance. And it is thus that I have seen fit to object that it is better, in this still pre-conversational moment, to speak instead of an intratext wherein contestation turns out to be confined to possibilities foreordained by uninterrogated hegemonic cosmological commitments. Michael Shapiro insists that "it must be recognized that the production of all texts (as well as their reading or consumption) involves acts of imagination" (Shapiro 1992: 17). Similarly, Darby proposes that imaginative literature is not as cleanly separable from international theory as has hitherto been imagined.[4] If all theory is made at least in part through acts of imagination, we are much the poorer for the foreclosure of possibilities that is effected by the hegemonologue even as we imagineer accounts of the international, or security, or the "good life." The Lakota idea of the sacred hoop is, among other things, a theory of the inter-national. It is also one from which we stand to learn a great deal. First, though, we will have to be willing to go where students and scholars of International Relations are not commonly found—places where their presence might, as it did at the McGill conference, raise a few well-disciplined eyebrows. And even as we venture to engage other voices, we must always bear in mind that the sites (and the minds) we occupy are not places we have in common. It is in these senses that an international relations conversation will be found(ed) in uncommon places. In the meantime, however, the hegemonologue speaks through us and our scholarship to secure the hegemonic knowledge system, the boundaries of our disciplines, and, no less, the contours and content of our key concepts and ideas. And speaking its voice, international theory (re)produces its exclusions and erasures, securing its own advanced colonial complicities.

Notes

1. Even poststructuralism, as I have argued, is limited by a de facto intratextuality— i.e., it is polyphonic only within the confines of a particular ethno-cultural milieu— and an incipient ethnocentrism that treads dangerously close to repudiating the possibility of an independent—that is, non-Western-derived—historical condition.
2. The Dull Knife breakout inspired John Ford's classic "Western" film, *Cheyenne Autumn*, starring Gilbert Roland and Ricardo Montalban.
3. As McDonald et al. put it "when history and archaeology are used by dominated groups, they can become tools capable of allowing the groups to free themselves from participation in the dominant ideology. . . . Both the archaeologists and the Northern Cheyenne learned that they can be natural allies, sometimes each possessing what the other needs" (McDonald et al. 1991: 77).
4. See Darby (1998).

Works Cited

Abu-Lughod, Lila, "Can There Be a Feminist Ethnography?" *Women & Performance* 5:1 (1990), pp. 7–27.

Albers, Patricia C., "Symbiosis, Merger, and War: Contrasting Forms of Intertribal Relationship Among Historic Plains Indians," in John H. Moore, ed., *The Political Economy of North American Indians* (Norman: University of Oklahoma Press, 1993).

Allen, Paula Gunn, *The Sacred Hoop: Recovering the Feminine in American Indian Traditions* (Boston: Beacon Press, 1992).

Alvarez-Pereyre, Frank, "From the Temptation for Purity to the Necessity of Unity: The Anthropological Sciences Put to the Test of Interdisciplinarity," *Diogenes* 159 (Winter 1992), pp. 95–135.

Amin, Samir, "1492," *Monthly Review* 44:3 (July–August 1992), pp. 10–19.

Anderson, Benedict, *Imagined Communities: Reflections on the Origin and Spread of Nationalism* (London: Verso, 1983).

Andrews, Lynn V., *Medicine Woman* (San Francisco: Harper & Row, 1981).

Anonymous, "Three Noted Chiefs of the Sioux," *Harper's Weekly* (October 20, 1890), p. 995.

Appadurai, Arjun, "Putting Hierarchy in Its Place," *Cultural Anthropology* 3:1 (February 1988), pp. 36–49.

———, *Modernity at Large: Cultural Dimensions of Globalization* (Minneapolis: University of Minnesota Press, 1996).

Appiah, Kwame Anthony, "Is the Post- in Postmodernism the Post- in Postcolonial?" *Critical Inquiry* 17:2 (Winter 1991), pp. 336–57.

Ashcroft, Bill, Gareth Griffiths, and Helen Tiffin, *The Empire Writes Back: Theory and Practice in Post-Colonial Literatures* (London: Routledge, 1989).

Ashley, Richard K., "The Geopolitics of Geopolitical Space: Toward a Critical Social Theory of International Politics," *Alternatives* 12:4 (October 1987), pp. 403–34.

Ashley, Richard K. and R.B.J. Walker, "Speaking the Language of Exile: Dissident Thought in International Studies," *International Studies Quarterly* 34:3 (September 1990a), pp. 259–68.

Ashley, Richard K. and R.B.J. Walker, "Reading Dissidence/Writing the Discipline: Crisis and the Question of Sovereignty in International Studies," *International Studies Quarterly* 34:3 (September 1990b), pp. 367–416.

Austin-Broos, Diane J., "Falling Through the 'Savage Slot': Postcolonial Critique and the Ethnographic Task," *Australian Journal of Anthropology* 9:3 (1998), pp. 295–309.

Bad Wound, Lewis, "The Sacred Hoop," in Roxanne Dunbar Ortiz, ed., *The Great Sioux Nation: Sitting in Judgment on America* (New York: American Indian Treaty Council Information Center, 1977).

Bamforth, Douglas B., "Indigenous People, Indigenous Violence: Precontact Warfare on the North American Great Plains," *MAN* 29:1 (March 1994), pp. 95–115.

Baran, Paul A., "The Commitment of the Intellectual," *Monthly Review* 13:1 (May 1961), pp. 8–18.

Basaglia, Franco, *Psychiatry Inside Out: Selected Writings of Franco Basaglia*, Nancy Scheper-Hughes and Anne M. Lovell, eds. (New York: Columbia University Press, 1987).

Baudrillard, Jean, *The Gulf War Did Not Take Place*, trans. Paul Patton (Bloomington: Indiana University Press, 1995).

Bedford, David and Thom Workman, "The Great Law of Peace: Alternative Inter-Nation(al) Practices and the Iroquoian Confederacy," *Alternatives* 22:1 (January–March 1997), pp. 87–111.

Bell, Diane, "Desperately Seeking Redemption," *Natural History* 106:2 (March 1997), pp. 52–53.

Bell, John, "The Sioux War Panorama and American Mythic History," *Theatre Journal* 48:3 (October 1996), pp. 279–99.

Berkhofer, Robert F., Jr., *The White Man's Indian: Images of the American Indian from Columbus to the Present* (New York: Vintage Books, 1978).

Bernal, Martin, *Black Athena: The Afroasiatic Roots of Classical Civilization* (New Brunswick: Rutgers University Press, 1987).

Bhabha, Homi, "Representation and the Colonial Text: A Critical Exploration of Some Forms of Mimeticism," in Frank Gloversmith, ed., *The Theory of Reading* (Brighton: Harvester Press, 1984).

———, "DissemiNation: Time, Narrative, and the Margins of the Modern Nation," in Homi K. Bhabha, ed., *Nation and Narration* (London: Routledge, 1990).

———, *The Location of Culture* (London: Routledge, 1994).

———, "Unpacking My Library . . . Again," in Iain Chambers and Lidia Curti, eds., *The Post-Colonial Question: Common Skies, Divided Horizons* (London: Routledge, 1996).

Biolsi, Thomas, "Ecological and Cultural Factors in Plains Indian Warfare," in R. Brian Ferguson, ed., *Warfare, Culture, and Environment* (Orlando: Academic Press, 1984).

———, "The IRA and the Politics of Acculturation: The Sioux Case," *American Anthropologist* 87:3 (September 1985), pp. 656–59.

———, "The Birth of the Reservation: Making the Modern Individual Among the Lakota," *American Ethnologist* 22:1 (February 1995), pp. 28–53.

————, "The Anthropological Construction of 'Indians': Haviland Scudder Mekeel and the Search for the Primitive in Lakota Country," in Thomas Biolsi and Larry J. Zimmerman, eds., *Indians and Anthropologists: Vine Deloria Jr. and the Critique of Anthropology* (Tucson: University of Arizona Press, 1997).

Bissonette, Gladys, "We Can Take Care of Our Own," in Roxanne Dunbar Ortiz, ed., *The Great Sioux Nation: Sitting in Judgment on America* (New York: American Indian Treaty Council Information Center, 1977).

Black Elk, Frank, "Observations on Marxism and Lakota Tradition," in Ward Churchill, ed., *Marxism and Native Americans* (Boston: South End Press, 1983).

Bleiker, Roland, "Discourse and Human Agency," *Contemporary Political Theory* 2:1 (March 2003).

Blick, Jeffrey P., "Genocidal Warfare in Tribal Societies as a Result of European-Induced Culture Conflict," *MAN* 23:4 (December 1988), pp. 654–70.

Boehmer, Elleke, *Colonial and Postcolonial Literature: Migrant Metaphors* (Oxford: Oxford University Press, 1995).

Bourgois, Philippe, "Confronting Anthropological Ethics: Ethnographic Lessons from Central America," *Journal of Peace Research* 27:1 (February 1990), pp. 43–54.

Brantlinger, Patrick, *Dark Vanishings: Discourse on the Extinction of Primitive Races, 1800–1930* (Ithaca: Cornell University Press, 2003).

Brown, Joseph Epes, *The Sacred Pipe: Black Elk's Account of the Seven Rites of the Oglala Sioux* (Norman: University of Oklahoma Press, 1953).

Buffalohead, W. Roger, "Reflections on Native American Cultural Rights and Resources," *American Indian Culture and Research Journal* 16:2 (1992), pp. 197–200.

Bunge, Robert, *An American Urphilosophie: An American Philosophy BP (Before Pragmatism)* (Lanham: University Press of America, 1984).

Butler, Judith, *Excitable Speech: A Politics of the Performative* (New York: Routledge, 1997).

Callender, Charles and Lee M. Kochems, "The North American Berdache," *Current Anthropology* 24:4 (August–October 1983), pp. 443–70.

Campbell, David, *Writing Security: United States Foreign Policy and the Politics of Identity* (Minneapolis: University of Minnesota Press, 1992).

————, "The Deterritorialization of Responsibility: Levinas, Derrida, and Ethics After the End of Philosophy," *Alternatives* 19:4 (Fall 1994), pp. 455–84.

————, "Violent Performances: Identity, Sovereignty, Responsibility," in Yosef Lapid and Friedrich Kratochwil, eds., *The Return of Culture and Identity in IR Theory* (Boulder: Lynne Rienner, 1996).

————, *National Deconstruction: Violence, Identity, and Justice in Bosnia* (Minneapolis: University of Minnesota Press, 1998).

Carr, E.H., *The Twenty Years Crisis, 1919–1939* (New York: Harper & Row, 1964).

Castile, George Pierre, "The Commodification of Indian Identity," *American Anthropologist* 98:4 (December 1996), pp. 743–49.

Chagnon, Napoleon A., *Yanomamö, the Fierce People* (New York: Holt, Rinehart and Winston, 1968).

Chagnon, Napoleon A., "Life Histories, Blood Revenge, and Warfare in a Tribal Population," *Science* 239 (February 26, 1988), pp. 985–92.

Chaloupka, William, *Knowing Nukes: The Politics and Culture of the Atom* (Minneapolis: University of Minnesota Press, 1992).

Chowdhry, Geeta and Sheila Nair, eds., *Power, Postcolonialism and International Relations: Reading Race, Gender and Class* (London: Routledge, 2002).

Churchill, Ward, *Fantasies of the Master Race: Literature, Cinema and the Colonization of American Indians* (Monroe: Common Courage Press, 1992).

———, *Indians Are Us? Culture and Genocide in Native North America* (Toronto: Between the Lines, 1994).

———, *Since Predator Came: Notes from the Struggle for American Indian Liberation* (Littleton: Aigis Publications, 1995).

Churchill, Ward and Winona LaDuke, "Native North America: The Political Economy of Radioactive Colonialism," in M. Annette Jaimes, ed., *The State of Native America: Genocide, Colonization, and Resistance* (Boston: South End Press, 1992).

Clark, Steve, "Introduction," in Steve Clark, ed., *Travel Writing and Empire: Postcolonial Theory in Transit* (London: Zed Books, 1999).

Clastres, Pierre, *Archeology of Violence*, trans. Jeanine Herman (New York: Semiotext(e), 1994).

Clifford, James, "Introduction: Partial Truths," in James Clifford and George E. Marcus, eds., *Writing Culture: The Poetics and Politics of Ethnography* (Berkeley: University of California Press, 1986a).

———, "On Ethnographic Allegory," in James Clifford and George E. Marcus, eds., *Writing Culture: The Poetics and Politics of Ethnography* (Berkeley: University of California Press, 1986b).

———, *Routes: Travel and Translation in the Late Twentieth Century* (Cambridge: Harvard University Press, 1997).

Cohn, Carol, "Sex and Death in the Rational World of Defense Intellectuals," *Signs: Journal of Women in Culture and Society* 12:4 (Summer 1987), pp. 687–718.

Cook, John R., *The Border and the Buffalo: An Untold Story of the Southwest Plains* (New York: Citadel Press, 1967).

Cook-Lynn, Elizabeth, "Who Gets to Tell the Stories?" *Wicazo Sa Review* 9:1 (Spring 1993), p. 9.

Cornell, Stephen, "The Transformations of Tribe: Organization and Self-Concept in Native American Ethnicities," *Ethnic and Racial Studies* 11:1 (January 1988), pp. 27–47.

Coser, Lewis A., *The Functions of Social Conflict* (New York: The Free Press, 1956).

Cox, Robert W., *Production, Power and World Order: Social Forces in the Making of History* (New York: Columbia University Press, 1987).

Craig, Robert, "Christianity and Empire: A Case Study of American Protestant Colonialism and Native Americans," *American Indian Culture and Research Journal* 21:2 (Spring 1997), pp. 1–41.

Crapanzano, Vincent, "Hermes' Dilemma: The Masking of Subversion in Ethnographic Description," in James Clifford and George E. Marcus, eds., *Writing*

Culture: The Poetics and Politics of Ethnography (Berkeley: University of California Press, 1986).

Crawford, James, ed., *Language Loyalties: A Source Book on the Official English Controversy* (Chicago: University of Chicago Press, 1992).

Crawford, Neta C., "A Security Regime Among Democracies: Cooperation Among Iroquois Nations," *International Organization* 48:3 (Summer 1994), pp. 345–85.

Dalby, Simon, "Security, Modernity, Ecology: The Dilemmas of Post–Cold War Security Discourse," *Alternatives* 17:1 (Winter 1992), pp. 95–134.

Darby, Phillip, "Postcolonialism," in Phillip Darby, ed., *At the Edge of International Relations: Postcolonialism, Gender & Dependency* (London: Pinter, 1997).

———, *The Fiction of Imperialism: Reading Between International Relations & Postcolonialism* (London: Cassell, 1998).

Darby, Phillip and A.J. Paolini, "Bridging International Relations and Postcolonialism," *Alternatives* 19:3 (Summer 1994), pp. 371–97.

Davies, J.C., "Euro-American Realism Versus Native Authenticity: Two Novels by Craig Lesley," *Studies in American Fiction* 22:2 (Autumn 1994), pp. 233–47.

Dawes, H.L., "Have We Failed the Indian?" *Atlantic Monthly* 84:502 (August 1899), pp. 280–85.

Deleuze, Gilles, *Foucault*, trans. Sean Hand (Minneapolis: University of Minnesota Press, 1988).

Deloria, Vine, Jr., *God is Red* (New York: Grosset & Dunlap, 1973).

———, "Circling the Same Old Rock," in Ward Churchill, ed., *Marxism and Native Americans* (Boston: South End Press, 1983).

———, "Commentary: Research, Redskins, and Reality," *American Indian Quarterly* 15:4 (Fall 1991), pp. 457–68.

———, "Is Religion Possible? An Evaluation of Present Efforts to Revive Traditional Tribal Religions," *Wicazo Sa Review* 8:1 (Spring 1992), pp. 35–39.

———, "Anthros, Indians, and Planetary Reality," in Thomas Biolsi and Larry J. Zimmerman, eds., *Indians and Anthropologists: Vine Deloria Jr. and the Critique of Anthropology* (Tucson: University of Arizona Press, 1997).

DeMallie, Raymond J., ed., *The Sixth Grandfather: Black Elk's Teachings Given to John G. Neihardt* (Lincoln: University of Nebraska Press, 1984).

———, "Lakota Belief and Ritual in the Nineteenth Century," in Raymond J. DeMallie and Douglas R. Parks, eds., *Sioux Indian Religion: Tradition and Innovation* (Norman: University of Oklahoma Press, 1987).

Denzin, Norman K., "The Epistemological Crisis in the Human Disciplines: Letting the Old Do the Work of the New," in Richard Jessor, Anne Colby, and Richard A. Shweder, eds., *Ethnography and Human Development: Context and Meaning in Social Inquiry* (Chicago: University of Chicago, 1996).

Der Derian, James, "The Boundaries of Knowledge and Power in International Relations," in James Der Derian and Michael J. Shapiro, eds., *International/Intertextual Relations: Postmodern Readings of World Politics* (Lexington: Lexington Books, 1989).

———, *Antidiplomacy: Spies, Terror, Speed and War* (Cambridge: Blackwell, 1992).

Der Derian, James, "Virtual Security: Technical Oversight, Simulated Foresight, and Political Blindspots in the Infosphere," in J. Marshall Beier and Steven Mataija, eds., *Cyberspace and Outer Space: Transitional Challenges for Multilateral Verification in the 21st Century: Proceedings of the Fourteenth Annual Ottawa NACD Verification Symposium* (Toronto: Centre for International and Security Studies, 1997).

Der Derian, James and Michael J. Shapiro, eds., *International/Intertextual Relations: Postmodern Readings of World Politics* (Lexington: Lexington Books, 1989).

Derrida, Jacques, *La Dissémination* (Paris: Éditions du Seuil, 1972).

———, *Speech and Phenomena and Other Essays on Husserl's Theory of Signs*, trans. David B. Allison (Evanston: Northwestern University Press, 1973).

———, *Of Grammatology*, trans. Gayatri Chakravorty Spivak (Baltimore: Johns Hopkins University Press, 1974).

———, *Positions*, trans. Alan Bass (Chicago: University of Chicago Press, 1981).

———, "Declarations of Independence," trans. Tom Keenan and Tom Pepper, *New Political Science* 15 (Summer 1986), pp. 7–15.

———, *The Post Card: From Socrates to Freud and Beyond*, trans. Alan Bass (Chicago: Univeristy of Chicago Press, 1987).

Detwiler, Fritz, "'All My Relatives': Persons in Oglala Religion," *Religion* 22:3 (July 1992), pp. 235–46.

Deudney, Daniel, "The Case Against Linking Environmental Degradation and National Security," *Millennium* 19:3 (Winter 1990), pp. 461–76.

Dogan, Mattei, "The New Social Sciences: Cracks in the Disciplinary Walls," *International Social Science Journal* 49:3 (September 1997), pp. 429–43.

Donaldson, Laura E., "On Medicine Women and White Shame-ans: New Age Native Americanism and Commodity Fetishism as Pop Culture Feminism," *Signs: Journal of Women and Culture in Society* 24:3 (Spring 1999), pp. 677–96.

Dorris, Michael, "Indians on the Shelf," in Calvin Martin, ed., *The American Indian and the Problem of History* (New York: Oxford University Press, 1987).

Doty, Roxanne Lynn, "Foreign Policy as Social Construction: A Post-Positivist Analysis of U.S. Counterinsurgency Policy in the Philippines," *International Studies Quarterly* 37:3 (September 1993), pp. 297–320.

Dunaway, Wilma A., "Incorporation as an Interactive Process: Cherokee Resistance to Expansion of the Capitalist-World System, 1560–1763," *Sociological Inquiry* 66:4 (November 1996), pp. 455–70.

Dunbar Ortiz, Roxanne, "Indian Political Economy," in Roxanne Dunbar Ortiz, ed., *The Great Sioux Nation: Sitting in Judgment on America* (New York: American Indian Treaty Council Information Center, 1977).

Durham, Jimmie, "Cowboys and . . . Notes on Art, Literature, and American Indians in the Modern American Mind," in M. Annette Jaimes, ed., *The State of Native America: Genocide, Colonization, and Resistance* (Boston: South End Press, 1992).

Dyc, Gloria, "The Use of Native Language Models in the Development of Critical Literacy," *American Indian Culture and Research Journal* 18:3 (1994), pp. 211–33.

Eco, Umberto, *Travels in Hyperreality*, trans. William Weaver (San Diego: Harcourt Brace & Company, 1986).

Engels, Friedrich, *The Origin of the Family, Private Property, and the State, in Light of the Researches of Lewis H. Morgan* (New York: International Publishers, 1972).

Enloe, Cynthia, *Bananas, Beaches, and Bases: Making Feminist Sense of International Relations* (Berkeley: University of California Press, 1989).

Epp, Roger, "At the Wood's Edge: Towards a Theoretical Clearing for Indigenous Diplomacies in International Relations," in D. Jarvis and R. Crawford, eds., *International Relations: Still an American Social Science?* (Albany: SUNY Press, 2000).

Ernest, Paul, *Social Constructivism as a Philosophy of Mathematics* (Albany: State University of New York Press, 1998).

Evans-Pritchard, E.E., *The Nuer: A Description of the Modes of Livelihood and Political Institutions of a Nilotic People* (New York: Oxford University Press, 1969).

Ewers, John C., "Intertribal Warfare as the Precursor of Indian-White Warfare on the Northern Great Plains," *Western Historical Quarterly* 6 (October 1975), pp. 397–410.

Fabian, Johannes, "Presence and Representation: The Other and Anthropological Writing," *Critical Inquiry* 16:4 (Summer 1990), pp. 753–72.

Fanon, Frantz, *Black Skin, White Masks*, trans. Charles Lam Markmann (New York: Grove Press, 1967).

Ferguson, R. Brian, "Blood of the Leviathan: Western Contact and Warfare in Amazonia," *American Ethnologist* 17:2 (May 1990), pp. 237–57.

———, "Tribal Warfare," *Scientific American* 266:1 (January 1992), pp. 108–13.

Ferguson, R. Brian and Neil L. Whitehead, "The Violent Edge of Empire," in R. Brian Ferguson and Neil L. Whitehead, eds., *War in the Tribal Zone: Expanding States and Indigenous Warfare* (Santa Fe: School of American Research Press, 1992).

Feyerabend, Paul K., *Against Method: Outline of an Anarchistic Theory of Knowledge* (London: NLB, 1975).

Fine, Gary Alan, "Ten Lies of Ethnography: Moral Dilemmas of Field Research," *Journal of Contemporary Ethnography* 22:3 (October 1993), pp. 267–93.

Forsyth, Murray, "Thomas Hobbes and the External Relations of States," *British Journal of International Studies* 5:3 (October 1979), pp. 196–209.

Foucault, Michel, *The Archaeology of Knowledge*, trans. A.M. Sheridan Smith (New York: Pantheon Books, 1972).

———, *Discipline and Punish: The Birth of the Prison*, trans. Alan Sheridan (New York: Vintage Books, 1977).

———, *The History of Sexuality*, vol.1, trans. Robert Hurley (New York: Vintage Books, 1978).

———, *Power/Knowledge: Selected Interviews and Other Writings, 1972–1977*, ed. Colin Gordon (New York: Pantheon Books, 1980).

Frankenberg, Ruth and Lata Mani, "Crosscurrents, Crosstalk: Race, 'Postcoloniality' and the Politics of Location," *Cultural Studies* 7:2 (May 1993), pp. 292–310.

Frederic J. Frommer, "Black Hills Are Beyond Price to Sioux," *Los Angeles Times* (August 19, 2001).

Friedman, Thomas L., *The Lexus and the Olive Tree* (New York: Farrar, Straus, Giroux, 1999).

Fukuyama, Francis, *The End of History and the Last Man* (New York: Free Press, 1992).

Fulton, Robert and Steven W. Anderson, "The Amerindian 'Man-Woman': Gender, Liminality, and Cultural Continuity," *Current Anthropology* 33:5 (December 1992), pp. 603–10.

Galaty, John G., "How Visual Figures Speak: Narrative Inventions of 'The Pastoralist' in East Africa," *Visual Anthropology* 15:3 (July 2002), pp. 347–67.

Galtung, Johan, "Violence, Peace, and Peace Research," *Journal of Peace Research* 6:3 (1969), pp. 167–91.

Geertz, Clifford, *The Interpretation of Cultures: Selected Essays* (New York: Basic Books, 1973).

———, *Local Knowledge: Further Essays in Interpretive Anthropology* (New York: Basic Books, 1983).

———, *Works and Lives: The Anthropologist as Author* (Stanford: Stanford University Press, 1988).

George, Jim, "Realist 'Ethics,' International Relations, and Post-modernism: Thinking Beyond the Egoism-Anarchy Thematic," *Millennium* 24:2 (Summer 1995), pp. 195–223.

Gilpin, Robert, "The Richness of the Tradition of Political Realism," *International Organization* 38:2 (1984), pp. 287–304.

———, "The Theory of Hegemonic War," in Robert I. Rotberg and Theodore K. Rabb, eds., *The Origin and Prevention of Major Wars* (Cambridge: Cambridge University Press, 1989).

Goetz, Anne Marie, "Feminism and the Limits of the Claim to Know: Contradictions in the Feminist Approach to Women in Development," *Millennium* 17:3 (Winter 1988), pp. 477–96.

Gose, Ben, "Indian Rituals, Jung and Nature Help Students Face Adulthood," *Chronicle of Higher Education* (December 6, 1996).

Greenberg, David F., "Why Was the Berdache Ridiculed?" *Journal of Homosexuality* 11:3–4 (Summer 1985), pp. 179–89.

Grim, John A., "Cultural Identity, Authenticity, and Community Survival: The Politics of Recognition in the Study of Native American Religions," *American Indian Quarterly* 20:3 (Summer 1996), pp. 353–76.

Grindstaff, Carl F., Wilda Galloway, and Joanna Nixon, "Racial and Cultural Identification Among Canadian Indian Children," *Phylon* 34:4 (December 1973), pp. 368–77.

Grinnell, George Bird, "The Indian on the Reservation," *Atlantic Monthly* 83:496 (February 1899), pp. 255–67.

Gross, Paul R. and Norman Levitt, *Higher Superstition: The Academic Left and Its Quarrels with Science* (Baltimore: Johns Hopkins University Press, 1994).

Hagedorn, Hermann, *Roosevelt in the Bad Lands* (Boston: Houghton Mifflin Company, 1930).

Hale, Charles R., "Travel Warning: Elite Appropriations of Hybridity, *Mestizaje*, Antiracism, Equality, and Other Progressive-Sounding Discourses in Highland Guatemala," *Journal of American Folklore* 112:445 (Summer 1999), pp. 297–315.

Hall, Stuart, "When was 'The Post-Colonial?' Thinking at the Limit," in Iain Chambers and Lidia Curti, eds., *The Post-Colonial Question: Common Skies, Divided Horizons* (London: Routledge, 1996).

Hannah, Matthew G., "Space and Social Control in the Administration of the Oglala Lakota ('Sioux'), 1871–1879," *Journal of Historical Geography* 19:4 (1993), pp. 412–32.

Harding, Sandra, *Is Science Multicultural? Postcolonialisms, Feminisms, and Epistemologies* (Bloomington: Indiana University Press, 1998).

Harrod, Howard L., *Becoming and Remaining a People: Native American Religions on the Northern Plains* (Tucson: University of Arizona Press, 1995).

Hartsock, Nancy, "Rethinking Modernism: Minority vs. Majority Theories," *Cultural Critique* 7 (Fall 1987), pp. 187–206.

Harvey, David, *The Condition of Postmodernity: An Enquiry into the Origins of Cultural Change* (Cambridge: Blackwell, 1990).

Hayward, Steve, "Against Commodification: Zuni Culture in Clarence Major's Native American Texts," *African American Review* 28:1 (Spring 1994), pp. 109–20.

Henderson, James (Sákéj) Youngblood, "The Context of the State of Nature," in Marie Battiste, ed., *Reclaiming Indigenous Voice and Vision* (Vancouver: UBC Press, 2000).

Hernández-Ávila, Inés, "Mediations of the Spirit: Native American Religious Traditions and the Ethics of Representation," *American Indian Quarterly* 20:3 (Summer 1996), pp. 329–52.

Hobbes, Thomas, *Leviathan* (New York: Penguin Books, 1968).

Hobson, Geary, "The Rise of the White Shaman as a New Version of Cultural Imperialism," in Geary Hobson, ed., *The Remembered Earth: An Anthology of Contemporary Native American Literature* (Albuquerque: University of New Mexico Press, 1979).

Hoffman, Stanley, *Janus and Minerva: Essays in the Theory and Practice of International Politics* (London: Westview Press, 1987).

Hoffman, Thomas J., "Moving Beyond Dualism: A Dialogue with Western European and American Indian Views of Spirituality, Nature, and Science," *The Social Science Journal* 34:4 (1997), pp. 447–60.

Holm, Tom, "The Militarization of Native America: Historical Process and Cultural Perception," *The Social Science Journal* 34:4 (1997), pp. 461–74.

Holsti, Ole R., "Models of International Relations and Foreign Policy," *Diplomatic History* 13:1 (Winter 1989), pp. 15–43.

Homan, Roger, "The Ethics of Open Methods," *British Journal of Sociology* 43:3 (September 1992), pp. 321–32.

Howard, Scott J., "Incommensurability and Nicholas Black Elk: An Exploration," *American Indian Culture and Research Journal* 23:1 (1999), pp. 111–37.

Hulme, Peter and Ludmilla Jordanova, "Introduction," in Peter Hulme and Ludmilla Jordanova, eds., *The Enlightenment and Its Shadows* (London: Routledge, 1990).

Hutcheon, Linda, " 'Circling the Downspout of Empire': Post-Colonialism and Postmodernism," *Ariel* 20:4 (October 1989), pp. 149–75.

Irigaray, Luce, *This Sex Which is Not One*, trans. Catherine Porter with Carolyn Burke (Ithaca: Cornell University Press, 1985a).

———, *Speculum of the Other Woman*, trans. Gillian C. Gill (Ithaca: Cornell University Press, 1985b).

Irwin, Lee, "Dreams, Theory, and Culture: The Plains Vision Quest Paradigm," *American Indian Quarterly* 18:2 (Spring 1994), pp. 229–45.

Jaimes, M. Annette, "Re-Visioning Native America: An Indigenist View of Primitivism and Industrialism," *Social Justice* 19:2 (Summer 1992), pp. 5–34.

Janis, Irving, *Groupthink: Psychological Studies of Policy Decisions and Fiascoes* (Boston: Houghton Mifflin, 1982).

Jenkins, Timothy, "Fieldwork and the Perception of Everyday Life," *MAN* 29:2 (June 1994), pp. 433–55.

Jervis, Robert, "Hypotheses on Misperception," *World Politics* 20:3 (April 1968), pp. 454–79.

Jocks, Christopher Ronwanièn:te, "Spirituality for Sale: Sacred Knowledge in the Consumer Age," *American Indian Quarterly* 20:3 (Summer 1996), pp. 415–31.

Johansen, Bruce and Roberto Maestas, *Wasi'chu: The Continuing Indian Wars* (New York: Monthly Review Press, 1979).

Johnson, Ronald M., "Schooling the Savage: Andrew S. Draper and Indian Education," *Phylon* 35:1 (March 1974), pp. 74–82.

Kant, Immanuel, *Perpetual Peace* (New York: Liberal Arts Press, 1948).

Kappler, Charles Joseph, comp. and ed., *Indian Affairs: Laws and Treaties*, vol.2 (New York: AMS Press, 1971).

Kasee, Cynthia R., "Identity, Recovery, and Religious Imperialism: Native American Women and the New Age," *Women & Therapy* 16:2–3 (Summer–Fall 1995), pp. 83–93.

Keal, Paul, *European Conquest and the Rights of Indigenous Peoples: The Moral Backwardness of International Society* (Cambridge: Cambridge University Press, 2003).

Keegan, John, "Warfare on the Plains," *The Yale Review* 84:1 (January 1996), pp. 1–48.

Keeley, Lawrence H., *War Before Civilization* (New York: Oxford University Press, 1996).

Kehoe, Alice B., "Primal Gaia: Primitivists and Plastic Medicine Men," in James A. Clifton, ed., *The Invented Indian: Cultural Fictions and Government Policies* (New Brunswick: Transaction Publishers, 1990).

Keohane, Robert O., *International Institutions and State Power: Essays in International Relations Theory* (Boulder: Westview Press, 1989).

Kills Straight, Birgil L., "We Can Take Care of Our Own," in Roxanne Dunbar Ortiz, ed., *The Great Sioux Nation: Sitting in Judgment on America* (New York: American Indian Treaty Council Information Center, 1977).

King, Cecil, "Here Come the Anthros," in Thomas Biolsi and Larry J. Zimmerman, eds., *Indians and Anthropologists: Vine Deloria Jr. and the Critique of Anthropology* (Tucson: University of Arizona Press, 1997).

King, Thomas, "Godzilla vs. Post-Colonial," *World Literature Written in English* 30:2 (1990), pp. 10–16.

Kingsbury, Benedict and Adam Roberts, "Introduction: Grotian Thought in International Relations," in Hedley Bull, Benedict Kingsbury, and Adam Roberts, eds., *Hugo Grotius and International Relations* (Oxford: Clarendon Press, 1990).

Klein, Bradley S., "After Strategy: The Search for a Post-Modern Politics of Peace," *Alternatives* 13:3 (July 1988), pp. 293–318.

Klein, Christina, " 'Everything of Interest in the Late Pine Ridge War Are Held By Us For Sale': Popular Culture and Wounded Knee," *American Historical Quarterly* 25:1 (Spring 1994), pp. 45–68.

Klein, Julie Thompson, "Interdisciplinary Needs: The Current Context," *Library Trends* 45:2 (Fall 1996), pp. 134–54.

Krause, Keith, "Critical Theory and Security Studies: The Research Programme of 'Critical Security Studies,' " *Cooperation and Conflict* 33:3 (September 1998), pp. 298–333.

Krishna, Sankaran, "The Importance of Being Ironic: A Postcolonial View on Critical International Relations Theory," *Alternatives* 18:3 (Summer 1993), pp. 385–417.

Krupat, Arnold, "Postcoloniality and Native American Literature," *Yale Journal of Criticism* 7:1 (1994), pp. 163–80.

Kulchyski, Peter, "From Appropriation to Subversion: Aboriginal Cultural Production in the Age of Postmodernism," *American Indian Quarterly* 21:4 (Fall 1997), pp. 605–20.

LaDuke, Winona, *All Our Relations: Native Struggles for Land and Life* (Cambridge: South End Press, 1999).

———, "Buffalo Nation," *Sierra* 85 (May/June 2000), pp. 66–73.

Lagrand, James B., "Whose Voices Count? Oral Sources and Twentieth-Century American Indian History," *American Indian Culture and Research Journal* 21:1 (1997), pp. 73–105.

Laxson, Joan D., "How 'We' See 'Them': Tourism and Native Americans," *Annals of Tourism Research* 18:3 (1991), pp. 365–91.

Lazarus, Edward, *Black Hills/White Justice: The Sioux Nation Versus the United States, 1775 to the Present* (New York: HarperCollins, 1991).

Lenin, Vladimir I., *Imperialism: The Highest Stage of Capitalism* (New York: International Publishers, 1939).

Levinas, Emmanuel, "The Trace of the Other," in Mark C. Taylor, ed., *Deconstruction in Context: Literature and Philosophy* (Chicago: University of Chicago Press, 1986).

Levy, Marc, "Is the Environment a National Security Issue?" *International Security* 20:2 (Fall 1995), pp. 35–62.

Lincoln, Bruce, "A Lakota Sun Dance and the Problematics of Sociocosmic Reunion," *History of Religions* 34:1 (August 1994), pp. 1–14.

Linker, Maureen, "Epistemic Relativism and Socially Responsible Realism: Why Sokal is not an Ally in the Science Wars," *Social Epistemology* 15:1 (2001), pp. 59–70.

Lizot, Jacques, "On Warfare: An Answer to N.A. Chagnon," *American Ethnologist* 21:4 (November 1994), pp. 845–62.

Locke, John, *Two Treatises of Government* (London: J.M. Dent and Sons, 1924).

Lonowski, Delmer, "A Return to Tradition: Proportional Representation in Tribal Government," *American Indian Culture and Research Journal* 18:1 (1994), pp. 147–63.

Loomba, Ania, *Colonialism/Postcolonialism* (London: Routledge, 1998).

Lugones, Maria and Elizabeth Spelman, "Have We Got a Theory for You! Feminist Theory, Cultural Imperialism and the Demand for 'The Woman's Voice'," *Women's Studies International Forum* 6:6 (Fall 1983), pp. 573–81.

Lyon, Arabella, "Interdisciplinarity: Giving Up Territory," *College English* 54:6 (October 1992), pp. 681–93.

Magnuson, Jon, "Selling Native American Soul," *Christian Century* 106:35 (November 22, 1989), pp. 1084–87.

Manzo, Kate, "Critical Humanism: Postcolonialism and Postmodern Ethics," *Alternatives* 22:3 (July 1997), pp. 381–408.

Maracle, Lee, "The 'Post-Colonial' Imagination," *Fuse* 16:1 (Fall 1992), pp. 12–15.

Marcus, George E., ed., *Critical Anthropology Now: Unexpected Contexts, Shifting Constituencies, Changing Agendas* (Santa Fe: School of American Research Press, 1999).

Marx, Karl, *The Eighteenth Brumaire of Louis Bonaparte* (New York: International Publishers, 1963).

Mathews, J. Tuchman, "The Environment and International Security," in Michael T. Klare and Daniel C. Thomas, eds., *World Security: Challenges for a New Century* (New York: St. Martin's Press, 1994).

Mayer, Frank H. and Charles B. Roth, *The Buffalo Harvest* (Denver: Sage Books, 1958).

McClintock, Anne, "The Angel of Progress: Pitfalls of the Term 'Post-colonialism'," in Patrick Williams and Laura Chrisman, eds., *Colonial Discourse and Post-Colonial Theory: A Reader* (New York: Columbia University Press, 1994).

McDonald, J. Douglas, Larry J. Zimmerman, A.L. McDonald, William Tall Bull, and Ted Rising Sun, "The Northern Cheyenne Outbreak of 1879: Using Oral History and Archaeology as Tools of Resistance," in R.H. McGuire and R. Paynter, eds., *The Archaeology of Inequality* (Oxford: Basil Blackwell, 1991).

McGinnis, Anthony, *Counting Coup and Cutting Horses: Intertribal Warfare on the Northern Plains, 1738–1889* (Evergreen: Cordillera Press, 1990).

McGuire, Randall H., "Why Have Archaeologists Thought the Real Indians Were Dead and What Can We Do About It?" in Thomas Biolsi and Larry J. Zimmerman, eds., *Indians and Anthropologists: Vine Deloria Jr. and the Critique of Anthropology* (Tucson: University of Arizona Press, 1997).

McKeon, Michael, "The Origins of Interdisciplinary Studies," *Eighteenth-Century Studies* 28:1 (Fall 1994), pp. 17–28.

Medicine, Bea, "The Interaction of Culture and Sex Roles in the Schools," *Integrateducation* 19:1–2 (January–April 1981), pp. 28–36.

Medicine, Beatrice, "Oral History," in Roxanne Dunbar Ortiz, ed., *The Great Sioux Nation: Sitting in Judgment on America* (New York: American Indian Treaty Council Information Center, 1977).

Melmer, David, "Politics Turns Ugly on Pine Ridge," *Indian Country Today* (May 10, 2000).

Metz, Sharon and Michael Thee, "Brewers Intoxicated With Racist Imagery," *Business and Society* 88 (Winter 1994), pp. 50–51.

Mohanty, Chandra Talpade, "Under Western Eyes: Feminist Scholarship and Colonial Discourses," *Boundary 2* 12:3/13:1 (Spring/Fall 1984), pp. 333–58.

Monture, Joel, "Native Americans and the Appropriation of Cultures," *Ariel* 25:1 (January 1994), pp. 114–21.

Morgan, Lewis Henry, *Ancient Society: Or Researches in the Lines of Human Progress From Savagery Through Barbarism to Civilization* (Chicago: C.H. Kerr, 1877).

Morgenthau, Hans, *Politics Among Nations: The Struggle for Power and Peace*, 6th edition (New York: Alfred A. Knopf, 1985).

Motohashi, Ted, "The Discourse of Cannibalism in Early Modern Travel Writing," in Steve Clark, ed., *Travel Writing and Empire: Postcolonial Theory in Transit* (London: Zed Books, 1999).

Mukherjee, Arun P., "Whose Post-Colonialism and Whose Post-Modernism?" *World Literature Written in English* 30:2 (1990), pp. 1–9.

Nagel, Joane and C. Matthew Snipp, "Ethnic Reorganization: American Indian Social, Economic, Political, and Cultural Strategies for Survival," *Ethnic and Racial Studies* 16:2 (April 1993), pp. 203–35.

Nandy, Ashis, *The Intimate Enemy: Loss and Recovery of Self under Colonialism* (Delhi: Oxford University Press, 1983).

Natrajan, Balmurli and Radhika Parameswaran, "Contesting the Politics of Ethnography: Towards an Alternative Knowledge Production," *Journal of Communication Inquiry* 21:1 (Spring 1997), pp. 27–59.

Neihardt, John G., *Black Elk Speaks: Being the Life Story of a Holy Man of the Oglala Sioux* (Lincoln: University of Nebraska Press, 1979).

Neufeld, Mark, *The Restructuring of International Relations Theory* (Cambridge: Cambridge University Press, 1995).

Neuffer, Elizabeth, "Germans Make a Hobby Out of Cowboys and Indians," *Boston Globe* (August 6, 1996).

Newton, Isaac, *Opticks: Or, A Treatise of the Reflections, Refractions, Inflections & Colours of Light* (New York: Dover Publications, 1952).

Nissani, Moti, "Ten Cheers for Interdisciplinarity: The Case for Interdisciplinary Knowledge and Research," *The Social Science Journal* 34:2 (1997), pp. 201–16.

Notarianni, Diane M., "Making Mennonites: Hopi Gender Roles and Christian Transformations," *Ethnohistory* 43:4 (Fall 1996), pp. 593–611.

Olson, William and Nicholas Onuf, "The Growth of a Discipline: Reviewed," in Steve Smith, ed., *International Relations: British and American Perspectives* (Oxford: Blackwell, 1985).

Oneal, John R. and Bruce Russett, "The Kantian Peace: The Pacific Benefits of Democracy, Interdependence, and International Organizations, 1885–1992," *World Politics* 52:1 (October 1999), pp. 1–37.

Ong, Aihwa, "Colonialism and Modernity: Feminist Re-presentations of Women in Non-Western Societies," *Inscriptions* 3/4 (1988), pp. 79–93.

Ostler, Jeffrey, "Conquest and the State: Why the United States Employed Massive Military Force to Suppress the Lakota Ghost Dance," *Pacific Historical Review* 65:2 (May 1996), pp. 217–48.

Owen, John M., "How Liberalism Produces Democratic Peace," *International Security* 19:2 (Fall 1994), pp. 87–125.

Parks, Douglas R. and Raymond J. DeMallie, "Plains Indian Native Literatures," *Boundary 2* 19:3 (Fall 1992), pp. 105–47.

Peterson, V. Spike, "Security and Sovereign States: What is at Stake in Taking Feminism Seriously?" in V. Spike Peterson, ed., *Gendered States: Feminist (Re)Visions of International Relations Theory* (Boulder: Lynne Rienner Publishers, 1992).

Powers, Marla N., "Menstruation and Reproduction: An Oglala Case," *Signs: Journal of Women in Culture and Society* 6:1 (Autumn 1980), pp. 54–65.

Powers, William K., *Oglala Religion* (Lincoln: University of Nebraska Press, 1975).

———, *Sacred Language: The Nature of Supernatural Discourse in Lakota* (Norman: University of Oklahoma Press, 1986).

———, "When Black Elk Speaks Everybody Listens," in Christopher Vecsey, ed., *Religion in Native North America* (Moscow: University of Idaho Press, 1990).

Pratt, Mary Louise, "Fieldwork in Common Places," in James Clifford and George E. Marcus, eds., *Writing Culture: The Poetics and Politics of Ethnography* (Berkeley: University of California Press, 1986).

———, "Arts of the Contact Zone," *Profession* 91 (1991), pp. 33–40.

———, *Imperial Eyes: Travel Writing and Transculturation* (London: Routledge, 1992).

Price, B. Byron, " 'Cutting for Sign': Museums and Western Revisionism," *Western Historical Quarterly* 24:2 (May 1993), pp. 229–34.

Price, Catherine, "Lakotas and Euroamericans: Contrasted Concepts of 'Chieftainship' and Decision-Making Authority," *Ethnohistory* 41:3 (Summer 1994), pp. 447–63.

Price, David H., "Cold War Anthropology: Collaborators and Victims of the National Security State," *Identities* 4:3–4 (June 1998), pp. 389–430.

Reddock, Rhoda, "(Post) Colonial Encounters of the Academic Kind: The National Security Question," *Identities* 4:3–4 (June 1998), pp. 467–74.

Rice, Julian, *Black Elk's Story: Distinguishing its Lakota Purpose* (Albuquerque: University of New Mexico Press, 1991).

Robbins, Rebecca L., "Self-Determination and Subordination: The Past, Present, and Future of American Indian Governance," in M. Annette Jaimes, ed., *The State of Native America: Genocide, Colonization, and Resistance* (Boston: South End Press, 1992).

Roos, Philip D., Dowell H. Smith, Stephen Langley, and James McDonald, "The Impact of the American Indian Movement on the Pine Ridge Reservation," *Phylon* 41:1 (March 1980), pp. 89–99.

Rosaldo, Renato, "From the Door of His Tent: The Fieldworker and the Inquisitor," in James Clifford and George E. Marcus, eds., *Writing Culture: The Poetics and Politics of Ethnography* (Berkeley: University of California Press, 1986).

Rose, Wendy, "The Great Pretenders: Further Reflections on Whiteshamanism," in M. Annette James, ed., *The State of Native America: Genocide, Colonization and Resistance* (Boston: South End Press, 1992).

Said, Edward, *Orientalism* (New York: Vintage Books, 1979).

———, *Culture and Imperialism* (New York: Vintage Books, 1993).

Scheper-Hughes, Nancy, "The Primacy of the Ethical: Propositions for a Militant Anthropology," *Current Anthropology* 36:3 (June 1995), pp. 409–20.

Schnarch, Brian, "Neither Man Nor Woman: Berdache—A Case for Non-Dichotomous Gender Construction," *Anthropologica* 34:1 (1992), pp. 105–21.

Secoy, Frank Raymond, *Changing Military Patterns on the Great Plains: 17th Century Through Early 19th Century* (Seattle: University of Washington Press, 1966).

Shanley, Kathryn W., "The Indians America Loves to Love and Read: American Indian Identity and Cultural Appropriation," *American Indian Quarterly* 21:4 (Fall 1997), pp. 675–702.

Shapiro, Michael J., *Reading the Postmodern Polity: Political Theory as Textual Practice* (Minneapolis: University of Minnesota Press, 1992).

Shaw, Karena, "Indigeneity and the International," *Millennium* 31:1 (2002), pp. 55–81.

Silko, Leslie Marmon, "An Old-Time Indian Attack Conducted in Two Parts," in Geary Hobson, ed., *The Remembered Earth: An Anthology of Contemporary Native American Literature* (Albuquerque: University of New Mexico Press, 1979).

Simon, William G. and Louise Spence, "Cowboy Wonderland, History, and Myth: 'It Ain't All That Different Than Real Life,'" *Journal of Film and Video* 47:1–3 (Spring–Fall 1995), pp. 67–81.

Smith, Andrea, "For All Those Who Were Indian in a Former Life," *Cultural Survival Quarterly* 17:4 (Winter 1994), pp. 70–71.

Smith, Linda Tuhiwai, *Decolonizing Methodologies: Research and Indigenous Peoples* (London: Zed Books, 1999).

Smith, Steve, "The Self-Images of a Discipline: A Genealogy of International Relations Theory," in Ken Booth and Steve Smith, eds., *International Relations Theory Today* (University Park: Pennsylvania State University Press, 1995).

Smithsonian Institution, "Showdown at 'The West as America' Exhibition," *American Art* 5:3 (Summer 1991), pp. 2–11.

Sokal, Alan D., "Transgressing Boundaries: Towards a Transformative Hermeneutics of Quantum Gravity," *Social Text* 46/47, 14:1–2 (Spring/Summer 1996a), pp. 217–52.

———, "A Physicist Experiments with Cultural Studies," *Lingua Franca* 6:4 (May/June 1996b), pp. 62–64.

Sokal, Alan D. and Jean Bricmont, *Impostures Intellectuelles* (Paris: Éditions Odile Jacob, 1997).

Sokal, Alan D. and Jean Bricmont, *Fashionable Nonsense: Postmodern Intellectuals' Abuse of Science* (New York: Picador, 1998).

Spivak, Gayatri Chakravorty, "Can the Subaltern Speak?" in Cary Nelson and Lawrence Grossberg, eds., *Marxism and the Interpretation of Culture* (Urbana: University of Illinois Press, 1988a).

———, *In Other Worlds: Essays in Cultural Politics* (New York: Routledge, 1988b).

———, "The *Intervention* Interview," with Terry Threadgold and Francis Bartkowski, *Southern Humanities Review* 22:4 (Fall 1988c), pp. 323–43.

Stacey, Judith, "Can There Be a Feminist Ethnography?" *Women's Studies International Forum* 11:1 (1988), pp. 21–27.

Standing Bear, Luther, *My People the Sioux* (Lincoln: University of Nebraska Press, 1975).

———, *Land of the Spotted Eagle* (Lincoln: University of Nebraska Press, 1978).

Stocking, George W., Jr., "Delimiting Anthropology: Historical Reflections on the Boundaries of a Boundless Discipline," *Social Research* 62:4 (Winter 1995), pp. 933–66.

Strong, Pauline Turner and Barrik Van Winkle, " 'Indian Blood': Reflections on the Reckoning and Refiguring of Native North American Identity," *Cultural Anthropology* 11:4 (November 1996), pp. 547–76.

Suganami, Hidemi, "Bringing Order to the Causes of War Debates," *Millennium* 19:1 (Spring 1990), pp. 19–35.

Sundstrom, Linea, "Smallpox Used Them Up: References to Epidemic Disease in Northern Plains Winter Counts, 1714–1920," *Ethnohistory* 44:2 (Spring 1997), pp. 303–43.

Sylvester, Christine, "Empathetic Cooperation: A Feminist Method for IR," *Millennium* 23:2 (Summer 1994), pp. 315–34.

Thayer, James Steel, "The Berdache of the Northern Plains: A Socioreligious Perspective," *Journal of Anthropological Research* 36:3 (Fall 1980), pp. 287–93.

Thin Elk, Marvin, "We Had No Choice," in Roxanne Dunbar Ortiz, ed., *The Great Sioux Nation: Sitting in Judgment on America* (New York: American Indian Treaty Council Information Center, 1977).

Tickner, J. Ann, "Hans Morgenthau's Principles of Political Realism: A Feminist Reformulation," *Millennium* 17:3 (Winter 1988), pp. 429–40.

———, *Gender in International Relations: Feminist Perspectives on Achieving Global Security* (New York: Columbia University Press, 1992).

Tinker, George E., *Missionary Conquest: The Gospel and Native American Cultural Genocide* (Minneapolis: Fortress Press, 1993).

Tohe, Laura, "There is No Word for Feminism in My Language," *Wicazo Sa Review* 15:2 (Fall 2000), pp. 103–10.

True, Jacqui, "Feminism," in Scott Burchill and Andrew Linklater et al., *Theories of International Relations* (New York: St. Martin's Press, 1996).

United States, Department of Justice, "Accounting for Native American Deaths, Pine Ridge Indian Reservation, South Dakota: Report of the Federal Bureau of Investigation, Minneapolis Division" (May 2000).

Utley, Robert M., *The Last Days of the Sioux Nation* (New Haven: Yale University Press, 1963).

Utley, Robert M. and Wilcomb E. Washburn, *Indian Wars* (Boston: Houghton Mifflin, 1977).

Van Der Dennen, J. and V. Falger, "Introduction," in J. Van Der Dennen and V. Falger, eds., *Sociobiology and Conflict: Evolutionary Perspectives on Competition, Cooperation, Violence and Warfare* (London: Chapman and Hall, 1990).

Venables, Robert, "Looking Back at Wounded Knee 1890," *Northeast Indian Quarterly* 7:1 (Spring 1990), pp. 36–37.

Walker, J.R., "The Sun Dance and Other Ceremonies of the Oglala Division of the Teton Dakota," *Anthropological Papers* 16 (1917), pp. 51–221.

Walker, R.B.J., *Inside/Outside: International Relations as Political Theory* (Cambridge: Cambridge University Press, 1993).

———, "The Subject of Security," in Keith Krause and Michael C. Williams, eds., *Critical Security Studies: Concepts and Cases* (Minneapolis: University of Minnesota Press, 1997).

Wallerstein, Immanuel, "What Are We Bounding, and Whom, When We Bound Social Research," *Social Research* 62:4 (Winter 1995), pp. 839–56.

Walt, Stephen, "The Renaissance of Security Studies," *International Studies Quarterly* 35:2 (June 1991), pp. 211–39.

———, "International Relations: One World, Many Theories," *Foreign Policy* 110 (Spring 1998), pp. 29–44.

Washburn, Wilcomb E., "A Fifty-Year Perspective on the Indian Reorganization Act," *American Anthropologist* 86:2 (June 1984), pp. 279–89.

———, "Response to Biolsi," *American Anthropologist* 87:3 (September 1985), p. 659.

Weatherford, J. McIver, *Indian Givers: How the Indians of the Americas Transformed the World* (New York: Crown Publishers, 1988).

Weightman, John, "The Lure of Unreason," *The Hudson Review* 51:3 (Autumn 1998), pp. 475–89.

Weltfish, Gene, "The Plains Indians: Their Continuity in History and Their Indian Identity," in Eleanor Burke Leacock and Nancy Oestreich Lurie, eds., *North American Indians in Historical Perspective* (New York: Random House, 1971).

White, Richard, "The Winning of the West: The Expansion of the Western Sioux in the Eighteenth and Nineteenth Centuries," *Journal of American History* 65:2 (September 1978), pp. 319–43.

White, Stephen K., *Sustaining Affirmation: The Strengths of Weak Ontology in Political Theory* (Princeton: Princeton University Press, 2000).

White Face, Charmaine, *Testimony for the Innocent* (Brunswick: Audenreed Press, 1998).

———, "Are you Lakota or Full Blood?" *Indian Country Today* (March 21, 2001).

Whitt, Laurie Anne, "Cultural Imperialism and the Marketing of Native America," *American Indian Culture and Research Journal* 19:3 (1995a), pp. 1–31.

———, "Indigenous Peoples and the Cultural Politics of Knowledge," in Michael K. Green, ed., *Issues in Native American Cultural Identity* (New York: Peter Lang, 1995b).

Whitworth, Sandra, *Feminism and International Relations: Towards a Political Economy of Gender in Interstate and Non-Governmental Institutions* (Houndsmills: Macmillan, 1994).

———, "The Practice, and Praxis, of Feminist Research in International Relations," in Richard Wyn Jones, ed., *Critical Theory and World Politics* (Boulder: Lynne Rienner, 2001).

Wight, Martin, "Why is There No International Theory?" in Herbert Butterfield and Martin Wight, eds., *Diplomatic Investigations: Essays in the Theory of International Politics* (Cambridge: Harvard University Press, 1966).

Willey, P. and Thomas E. Emerson, "The Osteology and Archaeology of the Crow Creek Massacre," *Plains Anthropologist* 38:145 (1993), pp. 227–69.

Williams, Walter L., "Persistence and Change in the Berdache Tradition Among Contemporary Lakota Indians," *Journal of Homosexuality* 11:3–4 (Summer 1985), pp. 191–200.

Wilmer, Franke, *The Indigenous Voice in World Politics: Since Time Immemorial* (Newbury Park: Sage, 1993).

———, "Indigenous Peoples, Marginal Sites, and the Changing Context of World Politics," in Francis A. Beer and Robert Hariman, eds., *Post-Realism: The Rhetorical Turn in International Relations* (East Lansing: Michigan State University Press, 1996).

Wilmer, Franke, Michael E. Melody, and Margaret Maier Murdock, "Including Native American Perspectives in the Political Science Curriculum," *PS: Political Science & Politics* 27:2 (June 1994), pp. 269–76.

Wilson, Angela Cavender, "American Indian History or Non-Indian Perceptions of American Indian History?" *American Indian Quarterly* 20:1 (Winter 1996), pp. 3–5.

Winch, Peter, *The Idea of a Social Science and its Relation to Philosophy*, 2nd ed. (London: Routledge, 1990).

Wolf, Eric R., *Europe and the People Without History* (Berkeley: University of California Press, 1982).

Wynne, Peter, "Return of the Native: David Carlson's *Dreamkeepers* Revives a Once-Popular American Opera Theme," *Opera News* 60:8 (January 6, 1996), pp. 28–29.

Zerubavel, Eviatar, "The Rigid, the Fuzzy, and the Flexible: Notes on the Mental Sculpting of Academic Identity," *Social Research* 62:4 (Winter 1995), pp. 1093–1106.

Zimmerman, Larry J. and Lawrence E. Bradley, "The Crow Creek Massacre: Initial Coalescent Warfare and Speculations About the Genesis of Extended Coalescent," *Plains Anthropologist* 38:145 (1993), pp. 215–26.

Index